AF413026

Endothelium-Derived Factors and Vascular Functions

Endothelium-Derived Factors and Vascular Functions

Proceedings of the Fourth International Symposium on Endothelium-Derived Factors, Tokyo, 7-9 December, 1993

Editor:

Tomoh Masaki
Department of Pharmacology
Kyoto University
Faculty of Medicine
Kyoto, Japan

 1994

EXCERPTA MEDICA,
Amsterdam, Lausanne, New York, Oxford, Shannon, Tokyo

International Congress Series No. 1051
ISBN 0-444-81669-0

This book is printed on acid-free paper.

Published by:
Elsevier Science B.V.
P.O. Box 211
1000 AE Amsterdam
The Netherlands

Library of Congress Cataloging in Publication Data:

```
International Symposium on Endothelium-Derived Factors (4th : 1993 :
  Tokyo, Japan)
    Endothelium-derived factors and vascular functions : proceedings
of the Fourth International Symposium on Endothelium-Derived
Factors, Tokyo, 7-9 December 1993 / editor, Tomoh Masaki.
        p.   cm. -- (International congress series ; no. 1051)
    Includes index.
    ISBN 0-444-81669-0 (alk. paper)
    1. Vascular endothelium--Molecular aspects--Congresses.
2. Endothelins--Congresses.  3. Nitric oxide--Physiological effect-
-Congresses.  4. Cell adhesion molecules--Congresses.  5. Cellular
signal transduction--Congresses.   I. Masaki, Tomoh.  II. Title.
III. Series.
    [DNLM: 1. Endothelium-Derived Relaxing Factor--congresses.
2. Vasoconstrictor Agents--congresses.  3. Vasodilator Agents-
-congresses.  4. Endothelium, Vascular--physiology--congresses.   W3
EX89 no. 1051 1994 / QV 150 I5914 1994e]
QP88.45.I58   1993
612.1'3--dc20
DNLM/DLC
for Library of Congress                                    94-16819
                                                              CIP
```

Printed in the Netherlands

Introduction

The title of this symposium, 'Endothelium-derived Factors and Vascular Functions', is truly a timely topic, being perhaps the most fascinating subject in biological and medical science today. When I was a medical student the histology professor taught us that the endothelium is simply a layer covering the inside of the vascular system, and virtually no physiological attention was paid to this tissue. This situation continued until quite recently. In fact a famous medical dictionary, published in 1990, still described the endothelium in this way.

In 1980, Professor R.F. Furchgott published a report in 'Nature' which completely changed our view of the endothelium. It overturned an important part of the traditional idea about the humoral control of the vascular system, an idea which had been shaped by respectable senior pharmacologists. A new era has begun and the factor, termed EDRF, has captured the fancy of so many physiologists and pharmacologists.

The 8th International Congress of Pharmacology was held in Tokyo in 1981, when we had the pleasure of having Professor Furchgott as a participant. We Japanese were fortunate to be able to hear the 'hottest' news from the discoverer himself!

The search for the nature of EDRF started immediately, and owing to the suggestion of Professor Furchgott, nitric oxide has become the strongest candidate for EDRF. In 1992 'Science' magazine called it the chemical of the year.

The discovery of EDRF naturally stimulated the effort of many scientists to identify EDCF (endothelium-derived constricting factor) from the endothelium . Some, including Professor P.M. Vanhoutte and the late Professor M. Fujiwara, have made a significant contribution to this project. In 1988 Professor T. Masaki and his colleagues published a paper in 'Nature' describing a new peptide called endothelin. Interestingly, two years before this, Masaki had shown no interest in the vascular system at all, except for its contractile proteins. This paper caused an explosive sensation in the scientific field and was, at the beginning of the 1990's, the most frequently cited work among all the biological articles.

The enthusiasm of scientists for endothelin has become greater and greater. This has encouraged studies on various aspects of the vascular system. The vascular system is perhaps now one of the biggest fields in biological science. There is no other area which has developed so suddenly, within such a short period of time.

Finally, I would like to emphasize once again, that this new current originated from the earnest, humble and dedicated effort of an old pharmacologist, using simple, classical equipment, who loves experimental research with the joy of a child.

Dr. Setsuro Ebashi
Member of the Japan Academy
Professor Emeritus, University of Tokyo
Professor Emeritus, National Institute for Physiological Sciences

Preface

This book is the published proceedings of the Uehara Memorial Foundation Symposium, held in Tokyo on December 7-9, 1993. Vascular biology is now a rapidly growing new field. The Uehara Memorial Foundation decided to promote the progress of this research field, and began to plan this meeting three years ago. Although this field had already expanded into a big field at that time, we decided to select topics in the symposium that focused on aspects such as endothelium derived vasoactive substances, particularly vasoconstrictors, vasodilators, and growth factors. Endothelial cells are known to play a pivotal role in controlling vasoconstriction and vasodilation, as well as vascular remodelling through production of these factors. In addition, they are important in vascular diseases, including vasospasm, hypertension and atherosclerosis. However, the regulatory mechanisms of the production of those factors in endothelial cells, and intracellular signalling stimulated by these factors in target cells, are still unclear. Therefore, this meeting was planned as an attempt to include these topics. In addition, we selected the topic on adhesion molecules related to the interaction of endothelial cells and blood cells, since there is now a major interest in this field.

Since this symposium was held in Tokyo, half of the speakers were from Japan. The other half consisted of a number of leading investigators from abroad. We think that this symposium was the most thorough and current presentation of the state of vascular biology in the world.

This area is still expanding. Whole aspects of this research field are still changing. Efforts should be directed to making the next breakthroughs. We hope that this book will encourage the reader in this direction.

Finally, we would like to express our cordial thanks to all the participants to this symposium, and to the Uehara Memorial Foundation for their kind and meticulous support.

Organizing Committee

Chairman:
Tomoh Masaki   Professor, Department of Pharmacology, Kyoto University, Faculty of Medicine, Kyoto, Japan

Members:
Hiroo Imura   President, Kyoto University, Kyoto, Japan
Sei-itsu Murota   Professor, Department of Physiological Chemistry, Graduate School, Tokyo Medical and Dental University, Tokyo, Japan
Shigetada Nakanishi   Professor, Institute for Immunology, Kyoto University, Faculty of Medicine, Kyoto, Japan
Fumimaro Takaku   President, International Medical Center of Japan, Japan

Contents

Introduction   v

Preface   vii

Endothelium-derived relaxing and constricting factors

Interactions of superoxide and hydrogen peroxide with nitric oxide and EDRF
in the regulation of vascular tone
   R.F. Furchgott, D. Jothianandan and N. Ansari   3

Regulation of vascular endothelial cell function by nitric oxide: Direct
inhibition of nitric oxide synthase
   L.J. Ignarro, G.M. Buga and J.M. Griscavage   13

Molecular biology of nitric oxide synthase and structure of endothelial nitric
oxide synthase gene
   Y. Yui, M. Kaoru, K. Sase, T. Kawamoto, K. Toda, Y. Doi, S. Ogoshi,
   R. Hattori, T. Aoyama, Y. Yamamoto, K. Hashimoto, Rie-XiA Yang,
   C. Kawai, S. Sasayama and Y. Shizuta   21

The nitric oxide-cyclic GMP signal transduction system
   F. Murad   29

Role of nitric oxide (NO) derived from endothelium and vasodilator nerve for
the regulation of cerebral arterial tone
   N. Toda   39

Endothelial NO and vascular regulation
   M. Félétou, E. Canet and P.M. Vanhoutte   47

Mechanisms of endothelin-induced contraction and relaxation of vascular
smooth muscle
   H. Kanaide   61

Endothelin receptor subtypes and mechanism of endothelin induced
vasoconstriction
   T. Masaki, A. Sakamoto and T. Enoki   71

Endothelin and renal tubules
   S. Uchida, F. Takemoto, K. Yoshitomi and K. Kurokawa   77

Establishment of mice carrying the mutant endothelin-1 gene by gene targeting
   Y. Kurihara, H. Kurihara, H. Suzuki, T. Kodama, K. Maemura, R. Nagai,
   H. Oda, T. Kuwaki, W.-H. Cao, N. Kamada, K. Jishage, Y. Ouchi,
   S. Azuma, Y. Toyoda, T. Ishikawa, M. Kumada and Y. Yazaki   87

Regulation of endothelin production by estrogen: effects of ovariectomy and
estrogen replacement in rats
   Y. Ouchi, M. Akishita, K. Kozaki, S. Kim, M. Ishikawa, K. Toba
   and H. Orimo   93

x

Vascular remodeling factors

Endothelial cell responses to fluid shear stress
    A. Kamiya, R. Korenaga and J. Ando ... 103
Possible participation of nitric oxide and endothelin-1 in the mechanism of
vascular remodeling through cell-to-cell interaction
    K. Fukuo, T. Inoue, T. Nabata, T. Nakahashi, S. Morimoto and T. Ogihara ... 113
C-type natriuretic peptide (CNP) as a novel endothelium-derived relaxing
peptide: Occurrence of vascular natriuretic peptide system
    K. Nakao, H. Itoh, T. Yoshimasa and H. Imura ... 123
Vasoactive factors in vascular remodeling: New insights from in vivo
gene transfer technology
    G.H. Gibbons, R.E. Pratt and V.J. Dzau ... 131

Growth factors in vascular cell function

Growth-regulatory molecules and atherosclerosis
    R. Ross ... 143
TGF-β: Structure and functional consequences
    D.B. Rifkin, C.N. Metz, I. Nunes and J. Harpel ... 151
Transforming growth factor-β receptors
    K. Miyazono and F. Takaku ... 157
Basic fibroblast growth factor (FGF-2) and potential mechanisms that regulate
its activities
    A. Baird ... 165
Coordinative effects of fibroblast growth factor and hypoxanthine on the
growth of porcine aortic endothelial cells
    A. Ichikawa, Y. Hayashi, S. Hirai, N. Nakanishi, T. Koizumi,
    Y. Morita and T. Fukui ... 171
Regulation of the cell cycle of vascular endothelial cells through the
protein kinase C pathway
    T. Sasaguri, C. Kosaka, K. Zen, J. Masuda, K. Shimokado and J. Ogata ... 177

Receptors and signal transduction in vascular systems

Signal transduction and ligand-binding domains of the tachykinin receptors
    S. Nakanishi, Y. Nakajima, Y. Yokota, U. Gether and T.W. Schwartz ... 191
Cloning, expression and regulation of angiotensin II receptors
    T. Inagami, N. Iwai, K. Sasaki, Y. Yamano, S. Bardhan, S. Chaki,
    D.-F. Guo, H. Furuta, K. Ohyama, Y. Kambayashi, K. Takahashi
    and T. Ichiki ... 199
Production of dominant negative mutations of guanylyl cyclase
    P.S.T. Yuen, M. Takada, D.K. Thompson and D.L. Garbers ... 205

Regulation of receptor-G protein signaling by the effector enzyme
phospholipase C-β
    E.M. Ross, G. Berstein and M. Karandikar   211
Phenotypic change of endothelin receptor subtype in vascular smooth muscle
cells: Its implication in atherosclerosis
    Y. Hirata and S. Eguchi   219

Adhesion molecules

Regulation of P-selectin function and expression
    R.P. McEver   227
Selectin-sialomucin interactions
    L.A. Lasky   233
An important role of adhesion molecules (LFA-1 & ICAM-1) in endothelial
cell injury caused by activated neutrophils
    S. Murota, H. Fujita, I. Morita and Y. Wakabayashi   241
Biological significance of L-selectin (LECAM-1) and its ligand
    M. Miyasaka, T. Tamatani and H. Kawashima   249
Novel monoclonal antibody (ASH1a/256C) recognizing asialo GM-2 developed
on the surface of fatty streak of atherosclerotic aorta
    T. Takano, M. Mori, K. Shima and T. Imanaka   253

Index of authors   257

Endothelium-derived relaxing and constricting factors

Endothelium-Derived Factors and Vascular Functions. T. Masaki, ed.

Interactions of superoxide and hydrogen peroxide with nitric oxide and EDRF in the regulation of vascular tone

Robert F. Furchgott, Desingarao Jothianandan and Nasrin Ansari

Department of Pharmacology, State University of New York Health Science Center at Brooklyn, Brooklyn NY 11203, USA

Abstract

Hydrogen peroxide (3–100 µM) produces relaxation of intact rings of rabbit aorta by facilitating release of EDRF. Xanthine plus xanthine oxidase (X + XO), a system generating superoxide, also produces relaxation mediated by the hydrogen peroxide produced by dismutation of superoxide. The superoxide from X + XO is a rapidly acting potent inhibitor of relaxation of rings by added NO. The immediate effectiveness of X + XO as an inhibitor of endothelium-dependent relaxation by acetylcholine is much greater when this superoxide generator is added a few minutes after acetylcholine rather than before or immediately after. This suggests that the nature of EDRF changes during the first few minutes of stimulated release.

Introduction

It is well established that the superoxide anion (O_2^-) and nitric oxide (NO) react extremely fast to generate peroxynitrite, which at physiological pH rapidly dissociates into a hydroxyl radical and NO_2, with the latter generating nitrite and nitrate ions [1,2]. Conversely, the enzyme superoxide dismutase (SOD) by catalyzing the dismutation of O_2^- to H_2O_2 and O_2, scavenges O_2^- that occurs in biological test systems, thus protecting added NO against inactivation and potentiating its vasorelaxing activity [2–6]. In perfusion-bioassay systems, endothelium-derived relaxing factor (EDRF) released from endothelial cells behaves similarly to NO with respect to inactivation by O_2^- and stabilization by SOD [6–9]. In organ chamber experiments with rings of rabbit thoracic aorta, we have frequently used the combination of xanthine (X) plus xanthine oxidase (XO) for generating O_2^- [2,5,6]. As briefly noted in a recent short paper [10], H_2O_2 derived from the dismutation of O_2^- generated by X plus XO in organ chamber experiments produces a relaxation of intact rings (rings with endothelium) of rabbit aorta. In that paper, it was shown that H_2O_2 itself in the range of 3–100 µM produces slow concentration-dependent relaxation of intact rings of aorta, and that this relaxation is due to H_2O_2-induced increase in release of EDRF from the endothelial cells. In the same paper [10], it was shown that H_2O_2 in the range of 1–100 µM inhibited in moderate degree and in a dose-dependent manner the transient relaxation of

4

endothelium-denuded aortic rings in response to added NO. Unexpectedly, this inhibition by H_2O_2 of NO-induced relaxation was markedly enhanced in the presence of added SOD. In view of the usual potentiation of NO-induced relaxation by SOD, this enhancement by SOD of inhibition by H_2O_2 was referred to as the "SOD paradox" [10].

In the present paper, we report the results of (1) more extensive experiments on endothelium-dependent relaxation induced by added H_2O_2 and by H_2O_2 produced by the dismutation of O_2^- generated by X plus XO; (2) additional experiments on the mechanism of the SOD paradox; and (3) a new series of experiments on factors influencing the effectiveness of X plus XO in inhibiting relaxation of intact aortic rings induced by acetylcholine (ACh). The results of this last series of experiments suggest that EDRF released from endothelial cells during continuous stimulation may change in nature during the first few minutes of stimulation.

Methods

The methods used to study the effects of various agents on contractile activity of rings of rabbit thoracic aorta mounted in organ chambers containing oxygenated Krebs solution were essentially the same as those previously described [2–6]. The rate of generation of O_2^- in Krebs solution in organ chambers from X plus XO (milk xanthine oxidase, Sigma) was determined by the rate of SOD-inhibitable reduction of oxidized cytochrome C pre-added to the chambers. In such experiments, catalase (200 U/ml) was also pre-added to prevent the accumulation of H_2O_2, which can interfere with the reactions. The order of addition of X and XO was optional, with zero time being the time when the second of the two components was added. Aliquots of solution removed at fixed intervals of time were immediately mixed with N-ethylmaleimide (NEM) to terminate the XO activity and then measured for absorbance changes at 550 nm in a recording spectrophotometer. To measure H_2O_2 concentrations at different time intervals after adding X plus XO to organ chambers (in the absence of cytochrome C and catalase), aliquots were removed, immediately mixed with NEM, and then reacted with dianisidine and horseradish peroxidase (HRP) and analyzed for absorbance at 500 nM.

Results and Discussion

Endothelium-dependent relaxation by H_2O_2

As reported previously [10], H_2O_2 in the range of 3 to 100 μM produced concentration-dependent relaxation of intact rings of rabbit aorta (Fig. 1). A typical response, as shown in the upper tracing of Fig. 1A, included an initial transient augmentation of tone followed by a slowly developing relaxation. At concentrations greater than 10 μM, H_2O_2 also produced a lesser degree of relaxation in matched endothelium-denuded rings (Fig. 1B). Denuded rings, unlike intact rings, gave no initial transient increase in tone (see [10]). On intact rings precontracted to about half-maximum with phenylephrine (PE), endothelium-dependent relaxation by H_2O_2 after correction for the endothelium-independent relaxation, averaged about 25 and 40% at 10 and 100 μM, respectively (Fig. 1B). Since this endothelium-dependent relaxation produced by H_2O_2 can be inhibited (or reversed) by catalase, hemoglobin (Hb) or N^G-monomethyl-L-arginine (L-NMMA) (as shown previously [10]), it is apparent that H_2O_2 stimulates release of EDRF.

When 10 μM H_2O_2 and a threshold dose of ACh (usually 10–20 nM) were added together on intact rings of aorta, there was a marked synergism, with the final endothelium-dependent-relaxation considerably exceeding the sum of the relaxing effects of the H_2O_2 and ACh alone (Fig. 1A, lower tracing). Synergism was of similar magnitude regardless of which agent was added first.

The endothelium-independent relaxation of rings of aorta produced by H_2O_2 at 100 μM and higher appeared to be the result of a direct irreversible depression of the contractile activity of the smooth muscle, since the sensitivity to contraction by PE remained depressed after removal of the H_2O_2 (not shown). When H_2O_2 was added at 200–300 μM for an extended period (20–30 min), it not only further depressed sensitivity to contraction by PE, but it now markedly impaired in an irreversible manner the endothelium-dependent relaxation of intact rings by ACh and by the calcium ionophone A23187 (not shown). Histological examination showed that this impairment was not due to loss of endothelial cells.

Hydrogen peroxide has been previously reported to relax rings of isolated arteries [11,12]. In the case of bovine pulmonary artery, no difference in degree of relaxation was noted in the absence and presence of endothelium [11], but in the case of canine coronary artery, a significantly greater degree of relaxation occurred in rings with endothelium [12]. In the present study on rings of rabbit aorta, there was also a greater relaxation in response to H_2O_2 when endothelium was present. With the use of selective inhibitors we have shown that this greater relaxation is the result of an increased release of EDRF. One possible explanation of this increased release is that H_2O_2 somehow interferes with mechanisms for removing free calcium

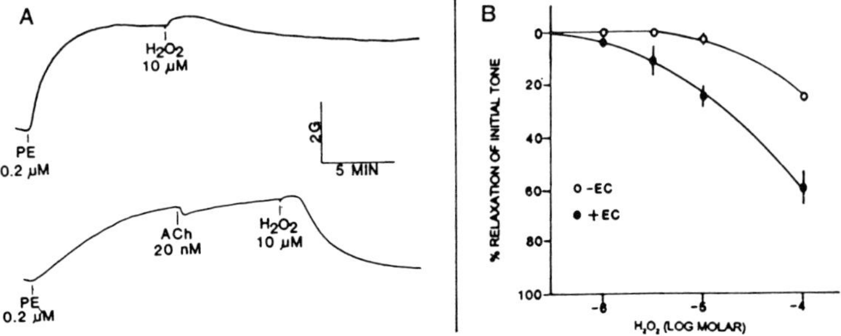

Fig. 1. Relaxation of rings of rabbit thoracic aorta by H_2O_2. **A**. Upper record: addition of 10 μM H_2O_2 to an intact ring produces a small slowly developing relaxation after an initial transient increase in contraction. Lower record: Synergism between threshold concentration of ACh (20 nM) and H_2O_2 (10 μM) in producing relaxation. **B**. Relaxation as function of H_2O_2 concentration for endothelium-intact (+EC) and endothelium-denuded (–EC) rings of aorta.

6

in endothelial cells, thus allowing a rise in cytoplasmic calcium concentrations which would increase the activity of constituitive nitric oxide synthase which synthesizes NO and citrulline from L-arginine and O_2. It has been reported that cultured endothelial cells exposed to H_2O_2 show an increase in cytoplasmic free calcium, as determined with fura-2 fluorescence [13]; however both the concentration of H_2O_2 (100 μM) and duration of exposure (1 hour or more) required to produce significant increases in endothelial cytoplasmic calcium were much larger than those required to increase EDRF release in our experiments on intact rings of rabbit aorta.

Effects of H_2O_2 and SOD on NO-induced relaxation: the SOD paradox

Figure 2 shows a record from an experiment of the type designed to quantify the inhibition of NO-induced relaxation by H_2O_2 in the presence and absence of SOD (35 U/ml). Cumulative additions of H_2O_2 (1–100 μM) on the control endothelium-denuded ring resulted in a moderate concentration-dependent inhibition (decrease in duration) of NO-induced relaxation. The usual potentiation by SOD alone of NO-induced relaxation is apparent from the much longer duration in the matched ring that had been pretreated with the enzyme. However, the potentiating effect did not persist in the presence of increasing concentrations of H_2O_2, and the SOD-treated ring actually exhibited a somewhat shorter duration of relaxation than the control at 10 μM H_2O_2 and a much shorter duration at 100 μM H_2O_2. This result is a manifestation of the "SOD paradox" [10]. Table 1 shows the results of two series of experiments of the type shown in Fig. 2. In the case of both series, one using 15 nM NO and the other 75 nM NO, the enhancement by SOD of the inhibition exerted by higher concentrations of H_2O_2 (10 and 100 μM) is apparent. The mechanism of the SOD paradox has not been determined with certainty, but our results favor the hypothesis that in the presence of high concentration of H_2O_2 and O_2 (the products of the dismutation reaction of O_2^-) the SOD catalysis of the reversal of the reaction yields O_2^- at a rate sufficient to cause a fairly rapid removal of NO by its reaction with the O_2^-.

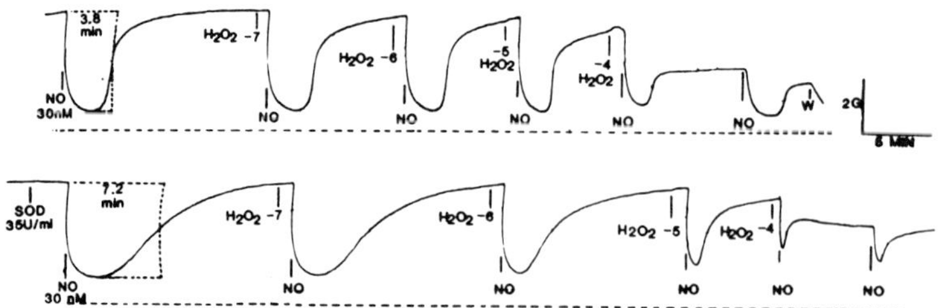

Fig. 2. The SOD paradox: inhibition of NO-induced relaxation by increasing concentrations of H_2O_2 is enhanced in the presence of SOD. Before addition of SOD (35 U/ml) to one of the two endothelium-denuded rings from the same aorta, they both had given relaxations of about the same duration (time from onset of relaxation to half-recovery of tone) in response to NO (30 nM). Dashed lines show basal tone levels before contraction with phenylephrine (PE), 0.1 μM.

Table 1. Inhibition by H_2O_2 of NO-induced relaxation of rabbit aorta in the presence and absence of SOD. Experimental design was similar to that shown in Fig. 2. Duration of relaxation was taken as the time from onset of relaxation on addition of NO to half-recovery of tone.

	Relative duration of relaxation in % of control			
H_2O_2 (µM)	0	1	10	100
Control, 15 nM NO	100*	80 ± 3	70 ± 5	42 ± 5
SOD, 15 nM NO	246 ± 13	129 ± 9	49 ± 1	20 ± 1
Control, 75 nM NO	100*	93 ± 3	74 ± 3	60 ± 4
SOD, 75 nM NO	134 ± 6	82 ± 4	34 ± 3	14 ± 2

* Control duration in absence of H_2O_2 and SOD was 2.99 ± 0.10 min for 15 nM NO and 7.04 ± 0.42 min for 75 nM NO (n = 4–13).

Endothelium-dependent relaxation by X plus XO

Relaxation of intact rings of aorta exposed to the O_2^--generating combination of X plus XO, and the enhancement of that relaxation in the presence of SOD is illustrated in Fig. 3. That such relaxation is the result of an action of H_2O_2 derived from the dismutation of the O_2^-, acting on the endothelial cells is evident from the ability of catalase to block (or reverse) the relaxation, and the absence of such relaxation in the case of endothelium-denuded rings, whether or not SOD was present (not shown). As would be expected on the basis of the studies on the endothelium-dependent relaxation by added H_2O_2 (see above), the relaxation of intact rings by X plus XO could be readily blocked (or reversed) by Hb or by L-NMMA (not shown), indicating that the generated H_2O_2 was stimulating the release of EDRF. As in the case of added H_2O_2, (Fig 1A, lower tracing) there was a marked synergism in producing relaxation when X plus XO, with SOD present, was added to intact aortic rings in the presence of a threshold concentration of ACh (not shown).

In experiments in which XO was added at 1.6 mU/ml and X at 0.1–0.3 mM to oxygenated Krebs solution in organ chambers, the rate of O_2^- generation (O_2^- flux) was about 5 µM/min. The dismutation of O_2^- to H_2O_2 in the absence as well as

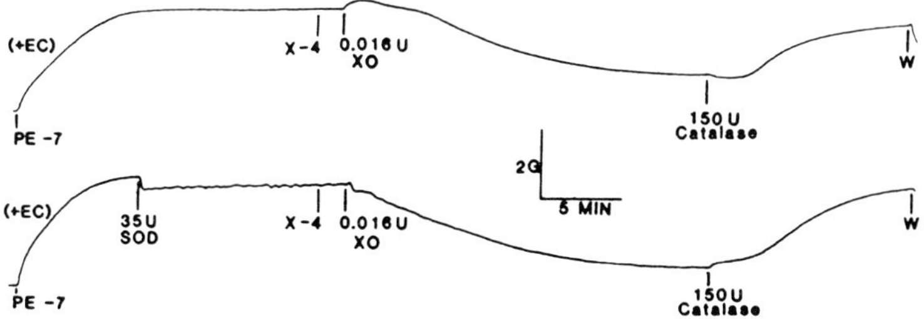

Fig. 3. Relaxation of intact rings of rabbit aorta by xanthine (X) plus xantine oxidase (XO): potentiation by SOD and reversal by catalase. Final concentrations in 20-ml organ chambers were PE, 0.1 µM; X, 0.1 mM; XO, 0.8 mU/ml; SOD, 35 U/ml; catalase, 150 U/ml.

presence of SOD resulted in the accumulation of H_2O_2 in the organ chambers, with the concentration reaching approximately 10 µM in the first 8 min after addition of X plus XO and a plateau of about 12 µM in 15 min. Thus, the concentration of H_2O_2 resulting from the oxidation of X by XO was of the magnitude that had been found to stimulate release of EDRF in the experiments with directly added H_2O_2 (see above).

Inhibition by X plus XO of relaxation induced by NO and EDRF

In the presence of freshly added X (0.1 mM) plus XO (0.8–1.6 mU/ml), the relaxation of endothelium-free rings produced by 50–150 nM NO (marked relaxation, lasting for several minutes) is completely blocked; and if the X plus XO is added during the course of maximum relaxation produced by the NO, the O_2^- generated by the combination rapidly and completely reverses the relaxation [6]. An example of reversal of NO relaxation is shown in Fig. 4A. Also shown in Fig. 4A is a typical example of the reversal by X plus XO of relaxation of an intact ring induced by ACh. Figure 4B illustrates the usefulness of having catalase present when testing the ability of X plus XO to reverse ACh-induced relaxation in intact rings. We began using catalase when we recognized that the H_2O_2 formed by the dismutation of O_2^- generated by X plus XO could itself produce endothelium-dependent relaxation (see above), and that this relaxation could interfere with the evaluation of the inhibition exerted by X plus XO against ACh-induced relaxation.

In a series of experiments with catalase (200 U/ml) present, the combination of X plus XO added 3 or more min after the development of full relaxation (usually 85–100% reduction of the initial phenylephrine-induced tone), rapidly reversed the relaxation. The reversal was complete (100% or greater) in all but 3 of the 14 rings. The mean degree of reversal was 114.0 ± 7.9%. In contrast to this excellent reversal of ACh-induced relaxation by X plus XO when this O_2^--generating system was added several minutes after the ACh, X plus XO added to the organ chamber prior to the addition of ACh only moderately inhibited the relaxation in response to the ACh (Fig. 5, left panel). In a large series of experiments the maximum relaxation produced by the addition of 1 µM ACh after the addition of X plus XO averaged about 65% of that in the absence of X plus XO and was reached about 2 min after the addition of ACh. However, following the attainment of maximum relaxation in response to ACh in the presence of X plus XO, the degree of inhibition of the ACh-induced relaxation gradually increased, so that after 10–20 min the final level of contraction was very similiar to that obtained at a similiar time after the acute reversal of maximal ACh-induced relaxation by the post-addition of X plus XO (Fig. 5).

When the combination of X plus XO was added after ACh but before the development of the full relaxing response to ACh, the rate of reversal of the relaxation became slower as the time was shortened between the addition of ACh and the addition of X plus XO (actually the addition of X with XO already present). This relationship between the time of the post-addition of X plus XO and the rate of reversal of ACh-induced relaxation is shown in Fig. 5 (right panel). The figure also shows how the addition of SOD (35 U/ml) rapidly restores the full relaxation by ACh, whether the blockade was the result of the pre- or post-addition of X plus XO. The ability of SOD to completely overcome blockade of ACh-induced relaxation by X plus XO, or if initially present, to completely prevent the occurrence of any blockade (not shown), clearly indicates that the blockade of ACh-induced relaxation by X plus XO is mediated by the O_2^- generated.

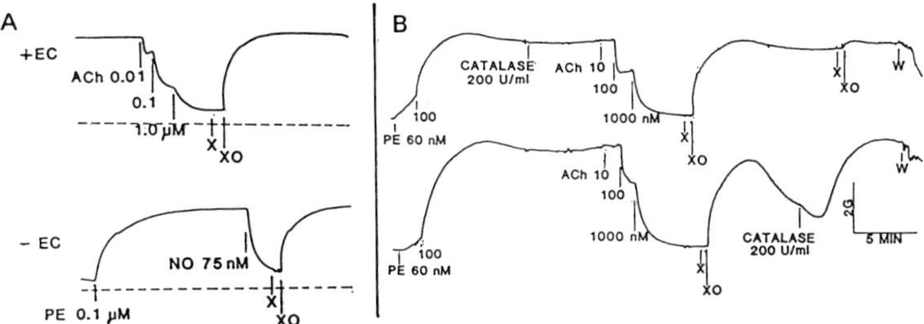

Fig. 4. Acute inhibition (rapid reversal) by xanthine (X) plus xanthine oxidase (XO) of relaxation of rabbit aortic rings induced by ACh and by NO, and influence of catalase on inhibition of ACh-induced relaxation. A. Simultaneous recordings of responses in an intact ring and an endothelium-denuded ring from the same aorta. Catalase (200 U/ml) had been pre-added to the intact ring. B. Simultaneous recordings of responses of two intact rings from another aorta. Inhibition of ACh-induced relaxation by X plus XO was sustained only in the ring pretreated with catalase. In the other ring, the initial inhibition decreased with time, but was then fully restored by the addition of catalase (X, 0.1 mM; XO, 1.6 mU/ml).

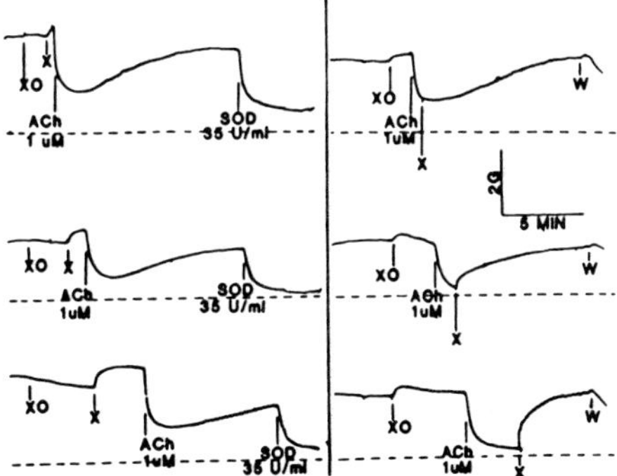

Fig. 5. Comparison of the rates of development of inhibition of ACh-induced relaxation in intact rings of rabbit aorta by xanthine plus xanthine oxidase added before and at different times after the addition of ACh. In this experiment with simultaneous recording from three matched rings from the same aorta, catalase (200 U/ml) was present throughout. After tests shown in left panel, rings were washed, and then contracted again with phenylephrine (0.1 μM) for tests shown in the right panel. The rate of development of inhibition was very much slower when the combination of X plus XO (to generate O_2^-) was added prior to or shortly after the addition of ACh than when it was added several minutes after ACh. The augmentation of contraction on pre-addition of X plus XO (left panel) is due to inhibition of basally released EDRF. SOD completely reestablished the ACh-induced relaxation in rings pretreated with X plus XO and in rings post-treated with X plus XO (not shown).

In a few experiments the calcium ionophore A23187 (3–10 nM) has been used in place of ACh to produce endothelium-dependent relaxation, and in those experiments also, pre-addition of X plus XO was much less effective than post-addition in producing rapid inhibition of the relaxation. In contrast to the discrepancy in initial effectiveness of pre- and post-additions of X plus XO in inhibiting relaxation of intact aortic rings by endothelium-dependent relaxing agents, there was no such discrepancy of effectiveness of pre- and post-additions of X plus XO in inhibiting endothelium-independent relaxation by added NO (50–150 nM) ([2,6] and unpubl. exp.).

At present, we are not certain why O_2^- generated by X plus XO appears to be less effective against endothelium-dependent relaxation when the X plus XO is added prior to ACh (or A23187). We have considered the possibility that if the released EDRF is simply NO, the initial rate of release immediately after addition of ACh (1 µM) might be so much greater than the rate a few minutes later that the flux of O_2^- generated by the pre-added X plus XO would be insufficient to block the large initial flux of EDRF. However, evidence against this was obtained in experiments in which ACh was added at very low concentrations (10–30 nM) giving only small relaxations (5–25% of maximum). Since these relaxations were also only partially inhibited by pre-addition of X plus XO (not shown), it is unlikely that an overwhelming initial flux of EDRF can account for the relatively poor blockade by the pre-added X plus XO.

At present, the hypothesis we favor is that the nature of EDRF released from the endothelial cells of rings of rabbit aorta changes during the first few minutes of stimulation by ACh (or A23187). According to this hypothesis, when ACh first begins acting on the cells, the EDRF released consists only in small part of NO and in large part of some compound of complex of NO (which will be referred to as RNO). RNO can also relax the smooth muscle cells, but unlike NO it is not readily inactivated by O_2^-. However, during the first few minutes of continued release, the fraction of RNO in the EDRF steadily decreases and that of NO increases, until the released EDRF is mainly NO. The hypothetical RNO might be a stored complex in limited amount, or it might be formed by the reaction of initially synthesized NO with some other cellular component, a limited supply of which could account for the shift in the composition of released EDRF from largely RNO to almost completely NO during the first few minutes of stimulation by ACh. Further experiments are needed to test this hypothesis and to identify RNO, if indeed it exists as another form of EDRF in addition to NO itself.

Acknowledgement

The research reported here was supported by USPHS Grant HL21860.

References

1. Beckman JS, Beckman JW, Chen J, Marshall PA. Proc Natl Acad Sci USA 1990;877:1620–1624.
2. Furchgott RF, Khan MT, Jothianandan D, In: Rubanyi GM, Vanhoutte PM (eds.) Endothelium derived relaxing factor. Basel: Karger, 1990;8–21.
3. Furchgott RF. In: Vanhoutte PM. (ed.) Vasodilatation: vascular smooth muscle, peptides, autonomic nerves and endothelium. New York: Raven Press, 1988;401–414.
4. Furchgott RF. Acta Physiol Scand 1990;139:257–270.
5. Furchgott RF, Jothianandan D, Freay AD. In: Moncada S, Higgs EA (eds.) Nitric oxide from L-arginine: a bioregulatory system. Amsterdam: Elsevier, 1990;5–17.
6. Furchgott RF, Jothianandan D, Khan MT. Jap J Pharmacol 1992;58, Suppl 2:185–191.
7. Moncada S, Palmer RMJ, Higgs EA. Pharmacol Rev 1991;43:109–142.
8. Khan MT, Furchgott RF. In: Rand MJ, Raper C (eds.) Additional evidence that endothelium derived relaxing factor is nitric oxide. Amsterdam: Elsevier, 1987;341–344.
9. Ignarro LJ, Byrns RE, Wood KS. In: Vanhoutte PM (ed.) Vasodilatation: vascular smooth muscle, peptides, autonomic nerves and endothelium. New York: Raven Press, 1988;427–435.
10. Furchgott RF, In: Mulvany M (ed.) Resistance arteries, structure and function. Amsterdam: Elsevier, 1991;216–220.
11. Rubanyi GM, Vanhoutte PM. Am J Physiol 1986;250:H815–H821.
12. Burke TM, Wolin MS. Am J Physiol 1987;252:H721–H732.
13. Kimura M, Maeda K, Hayashi S. Br J Pharmacol 1992;107:488–493.

Endothelium-Derived Factors and Vascular Functions. T. Masaki, ed.

Regulation of vascular endothelial cell function by nitric oxide: direct inhibition of nitric oxide synthase

Louis J. Ignarro, Georgette M. Buga and Jeanette M. Griscavage

Department of Pharmacology, UCLA School of Medicine; and Center for the Health Sciences, Los Angeles, CA 90024, USA

Summary

The objective of the present study was to determine the mechanism by which nitric oxide (NO) inhibits NO synthase and whether NO can inhibit endothelium-dependent vasorelaxation mediated by activation of NO synthase. NO inhibited the constitutive isoforms of NO synthase from rat cerebellum and bovine aortic endothelial cells. Purified NO synthase was much more sensitive than unpurified enzyme to the inhibitory effect of NO. Purified NO synthase was inhibited by NO by a direct interaction between NO and enzyme-bound heme. Inhibition of NO synthase by NO was observed also in intact bovine pulmonary arterial rings. Pretreatment of arterial rings with NO attenuated endothelium-dependent but not independent vasorelaxation. Isolated aortic endothelial cells perfused in a bioassay cascade generated NO in response to shear stress and bradykinin, which was markedly inhibited by pretreatment of cells with NO. These observations indicate that NO can modulate endothelium-dependent vasorelaxation by a negative feedback effect on NO synthase in the L-arginine-NO pathway.

Introduction

Previous studies from this laboratory revealed that NO could inhibit NO synthase prepared from the cytosolic fraction of rat cerebellum [1]. NO donor compounds as well as free NO inhibited NO synthase activity as assessed by measurement of citrulline formation from L-arginine. The non linearity of product formation under incubation conditions of saturating concentrations of substrate and cofactors suggested that negative feedback product inhibition might be occurring. NO but not citrulline, nitrite or nitrate inhibited NO synthase activity. In addition to the constitutive isoform of NO synthase from rat cerebellum, the inducible isoform of NO synthase from cytokine-activated macrophages was also inhibited by NO [2,3], although the inducible enzyme was much less sensitive than the constitutive enzyme to the inhibitory action of NO. Like the constitutive isoform from rat cerebellum, the constitutive membrane-bound isoform of NO synthase from aortic endothelial cells was very sensitive to the inhibitory action of NO [4].

The objective of the present study was to elucidate the mechanism by which NO inhibits the constitutive isoform of NO synthase. Moreover, experiments were conducted to determine whether NO could attenuate NO synthase activity in intact

14

vascular preparations including endothelial cells and thereby inhibit endothelium-dependent, NO-mediated vasorelaxation.

Experimental procedures

NO synthase assay

NO synthase activity was determined by measuring the formation of citrulline from L-arginine as described previously [5]. The supernatant fraction obtained by centrifugation (100,000 × *g* for 60 min) of 25% (w/v) homogenates of rat cerebellum was employed as the source of unpurified neuronal NO synthase [5]. Neuronal NO synthase was purified to homogeneity from the supernatant fraction by sequential column chromatography on DEAE cellulose followed by 2′,5′-ADP sepharose affinity resin. Endothelial NO synthase was prepared as a washed crude membrane fraction from cultured bovine aortic endothelial cells as described previously [4].

Assay reaction mixtures for NO synthase consisted of 50 mM Tris HCl, pH 7.4, 50 µM L-arginine (200,000 dpm of L-[2,3,4,5-^{3}H]arginine HCl, 77 Ci/mmol), 100 µM NADPH, 10 µM FAD, 10 µM tetrahydrobiopterin, 1 µg calmodulin and 2 mM $CaCl_2$ in a final incubation volume of 0.1 ml. The concentrations of enzyme fraction proteins were 0.24 mg (unpurified neuronal NO synthase), 0.62 mg (unpurified endothelial NO synthase), and 0.26 µg (purified neuronal NO synthase). Incubations were conducted at 37°C for 10–15 min. Reactions were terminated and samples were chromatographed as described previously [5].

Vasorelaxation experiments

Rings of bovine intrapulmonary artery were prepared and mounted under optimal tension for recording of changes in isometric force as described previously [4,6]. Rings were precontracted by addition of 1–10 µM phenylephrine and were washed and allowed to equilibrate for 30 min in between additions of drugs to obtain smooth muscle responses.

Results and Discussion

Inhibition of NO synthase activity by NO

Experiments with unpurified preparations of cytosolic constitutive NO synthase from rat cerebellum revealed that NO (10–50 µM) and the NO donor compound *S*-nitroso-*N*-acetylpenicillamine (SNAP; 50–500 µM) inhibited citrulline formation by NO synthase to the extent of 25–90%. The inhibitory action of NO and SNAP was enhanced about 2-fold by 200 units/ml of superoxide dismutase and abolished by 30 µM oxyhemoglobin (Table 1). Similarly, NO synthase activity in the absence

Table 1. Inhibitory action of NO on neuronal NO synthase

Additions		NO synthase activity[a] (pmol L-citrulline/min/mg protein)		
		Control	+100 U SOD	+30 µM HbO$_2$
None		362 ± 21	291 ± 18	483 ± 31
NO	10 µM	286 ± 17	143 ± 9	516 ± 43
	50 µM	60 ± 3	0	336 ± 27
SNAP	50 µM	264 ± 21	147 ± 12	506 ± 29
	200 µM	175 ± 12	88 ± 6	487 ± 34

[a]Standard incubations were conducted for 15 min as described in "Experimental Procedures". Superoxide dismutase (SOD) and oxyhemoglobin (HbO$_2$) were added to reaction mixtures prior to, whereas NO and S-nitroso-N-acetylpenicillamine (SNAP) were added after, addition of NO synthase. Values are expressed as the mean ± S.E.M. of duplicate determinations from four separate experiments.

of added NO or SNAP was inhibited about 30% by superoxide dismutase and enhanced about 2-fold by oxyhemoglobin. In the presence of saturating concentrations of L-arginine substrate and the cofactors NADPH, FAD, tetrahydrobiopterin, and calmodulin, citrulline formation catalyzed by NO synthase was nonlinear over an incubation interval at 37°C of 15 min. The addition of superoxide dismutase to enzyme reaction mixtures made the reaction rate even more nonlinear, whereas addition of oxyhemoglobin made the reaction rate linear. These observations indicate that added NO inhibits NO synthase and that enzymatically generated NO also inhibits NO synthase activity. Thus, enzymatically formed NO may elicit a negative feedback effect on NO synthase and thereby control its own biosynthesis.

The inhibitory effect of superoxide dismutase on NO synthase and its capacity to enhance the inhibitory action of NO may be attributed to the dismutation or destruction of superoxide anion, thereby prolonging the half-life of NO. Superoxide anion is known to react with NO to form peroxynitrite anion [7,8]. Inhibition of this reaction would be expected to prolong the half-life of NO [9], and thereby enhance the inhibitory action of NO on NO synthase. If NO elicits a negative feedback effect on NO synthase, then the addition to enzyme reaction mixtures of a compound that sequesters an/or destroys NO should increase enzymatic activity. Oxyhemoglobin is well-known to bind and oxidize NO, and this compound abolished the inhibitory action of NO on NO synthase.

The membrane-bound, constitutive isoform of NO synthase was prepared from cultured bovine aortic endothelial cells and examined for the influence of NO and NO donor compounds on enzymatic activity. Table 2 reveals that NO and SNAP both inhibited NO synthase in a concentration-dependent manner. Superoxide dismutase inhibited NO synthase activity and enhanced the inhibitory action of NO. Oxyhemoglobin enhanced NO synthase activity and abolished the inhibitory action of NO. Thus, the endothelium-derived NO synthase behaved similarly to the neuronally-derived NO synthase with regard to the inhibitory action of NO.

16

Table 2. Inhibitory action of NO on endothelial NO synthase

Additions		NO synthase activity[a] (pmol L-citrulline/min/mg protein)		
		Control	+100 U SOD	+30 µM HbO$_2$
None		24 ± 1.6	20 ± 1.4	34 ± 2.3
NO	10 µM	16 ± 1.1	10 ± 1.1	36 ± 3.1
	100 µM	9 ± 0.8	4 ± 0.5	32 ± 2.1
SNAP	200 µM	7 ± 0.5	3 ± 0.2	32 ± 2.7
	400 µM	3 ± 0.4	0	31 ± 1.9

[a]Standard incubations were conducted for 15 min as described in "Experimental Procedures". Superoxide dismutase (SOD) and oxyhemoglobin (HbO$_2$) were added to reaction mixtures prior to, whereas NO and S-nitroso-N-acetylpenicillamine (SNAP) were added after, addition of NO synthase. Values are expressed as the mean ± S.E.M. of duplicate determinations from three separate experiments.

The findings that NO inhibited unpurified preparations of NO synthase could not be interpreted as a direct effect on NO synthase as indirect mechanisms could not be ruled out. NO synthase purified from rat cerebellum to a very high specific activity was very sensitive to the inhibitory effect of NO and SNAP. Table 3 illustrates that NO and SNAP at concentrations of 0.1–10 µM inhibited NO synthase. The purified NO synthase was approximately 100-fold more sensitive than the unpurified enzyme to the inhibitory action of NO. The fact that NO inhibited the purified homogeneous enzyme signifies that a direct interaction between NO and NO synthase accounts for the negative effect of NO. The observations made with superoxide dismutase and oxyhemoglobin (Table 3) indicate that enzymatically generated NO from L-arginine inhibits NO synthase activity.

Although the mechanism of inhibitory action of NO is unknown, it appears likely that NO binds to the heme prosthetic group of NO synthase and thereby interferes with the oxygenation of L-arginine to its products NO and citrulline. NO synthase

Table 3. Inhibitory action of NO on purified neuronal NO synthase

Additions		NO synthase activity[a] (nmol L-citrulline/min/mg protein)		
		Control	+20 U SOD	+30 µM HbO$_2$
None		543 ± 32	426 ± 18	964 ± 47
NO	0.1 µM	481 ± 26	204 ± 11	981 ± 53
	1 µM	209 ± 18	89 ± 7	940 ± 37
SNAP	1 µM	453 ± 28	257 ± 22	918 ± 40
	10 µM	227 ± 21	108 ± 8	931 ± 51

[a]Standard incubations were conducted for 10 min as described in "Experimental Procedures". Superoxide dismutase (SOD) and oxyhemoglobin (HbO$_2$) were added to reaction mixtures prior to, whereas NO and S-nitroso-N-acetylpenicillamine (SNAP) were added after, addition of NO synthase. Values are expressed as the mean ± S.E.M. of duplicate determinations from three to four separate experiments.

is known to be a hemoprotein [10–13] and NO has a very high binding affinity for hemoproteins. Thus, it is likely that NO competes with oxygen for binding sites on enzyme-bound heme. This interpretation is consistent with the observation that added oxyhemoglobin abolished the inhibitory effect of NO.

Inhibition of endothelium-dependent vasorelaxation by NO

A recent study from this laboratory revealed that NO in the form of the NO donor compound SNAP inhibited endothelium-dependent vasorelaxation of bovine pulmonary arterial rings caused by acetylcholine and bradykinin, but did not alter endothelium-independent vasorelaxation caused by SNAP itself [4]. In the present study, *N*-methyl-*N'*-nitro-*N*-nitrosoguanidine (MNNG) was used as the NO donor drug. Figure 1 illustrates that pretreatment of arterial rings with MNNG resulted in the attenuation of endothelium-dependent vasorelaxation elicited by acetylcholine

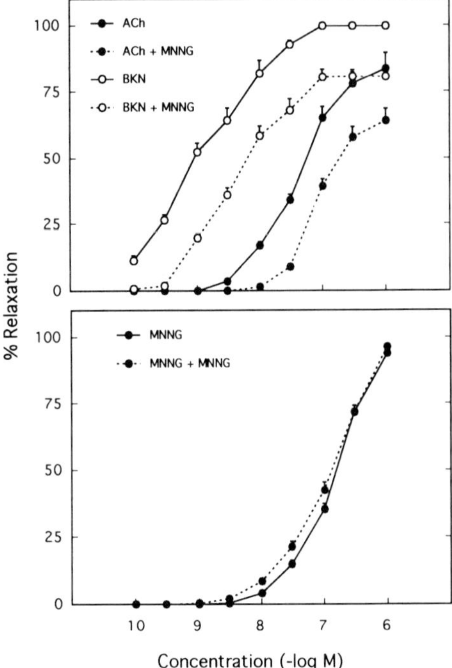

Fig. 1. Inhibition of endothelium-dependent but not endothelium-independent relaxation of isolated bovine pulmonary arterial rings by NO in the form of *N*-methyl-*N'*-nitro-*N*-nitrosoguanidine (MNNG). Control relaxant responses to bradykinin (BKN), acetylcholine (ACh), or MNNG were obtained, and the rings were washed three times. After a 15 min interval, the rings were washed extensively over a 40 min period. A second set of control responses to BKN, ACh, or MNNG was obtained, the tissues were washed three times, and then the arterial rings were bathed in 1 µM MNNG for 15 min. After extensive washing of the rings over a 40 min period, responses to added BKN, ACh, or MNNG were obtained. Responses to BKN and ACh are illustrated in the upper graph, whereas responses to MNNG are illustrated in the lower graph. Values represent the mean ± S.E.M. of 12 determinations from four separate experiments.

18

and bradykinin. In contrast, pretreatment of arterial rings with MNNG failed to alter subsequent relaxant responses to MNNG itself, an endothelium-dependent vasorelaxant that produces its effects by discharging NO [14].

These findings indicate that NO inhibits endothelium-dependent vasorelaxation but do not provide information regarding mechanisms. Another study from this laboratory [4], however, revealed that NO and SNAP can markedly diminish or abolish the formation of NO by perfused aortic endothelial cells. Thus, the likely mechanism by which NO attenuates endothelium-dependent vasorelaxation is inhibition of endothelial NO synthase.

In vivo studies are in progress to determine whether NO can inhibit endothelium-dependent vasodilation. A hindquarters perfusion circuit was set up in anesthetized rabbits under conditions of constant blood flow so that changes in hindquarters perfusion pressure are a reflection of changes in vascular resistance. SNAP was infused (100 µg/kg/min) into the hindquarters perfusion circuit together with sufficient phenylephrine to maintain a constant and normal baseline perfusion pressure. The infusion of SNAP markedly inhibited endothelium-dependent vasodilation elicited by bolus injections of acetylcholine or bradykinin into the hindquarters perfusion circuit without significantly altering vasodilator responses to injected NO.

Collectively, these observations indicate that NO inhibits the constitutive isoforms of NO synthase by mechanisms involving an interaction with the heme iron bound to NO synthase. The inhibition of NO synthase by NO appears to be physiologically important because enzymatically generated NO inhibits NO synthase and pretreatment of isolated arterial rings and endothelial cells with NO markedly interferes with endothelium-dependent, NO-mediated vasorelaxation.

Acknowledgements

This study was supported by National Institutes of Health Grants HL40922 and HL35014, the Laubisch Fund for Cardiovascular Research, and the Tobacco-Related Disease Research Program.

References

1. Rogers NE, Ignarro LJ. Biochem Biophys Res Commun 1992;189:242–249.
2. Griscavage JM, Rogers NE, Sherman MP, Ignarro LJ. J Immunol (in press).
3. Assreuy J, Cunha IQ, Liew FY, Moncada S. Br J Pharmacol 1993;108:833–837.
4. Buga GM, Griscavage JM, Rogers NE, Ignarro LJ. Circ Res 1993;73:808–812.
5. Bush PA, Gonzalez NE, Griscavage JM, Ignarro LJ. Biochem Biophys Res Commun 1992;185: 960–966.
6. Ignarro LJ, Byrns RE, Wood KS. Circ Res 1987;60:82–92.
7. Blough NV, Zafiriou OC. Inorg Chem 1985;24:3502–3504.
8. Beckman JS, Beckman TW, Chen J, Marshall PA, Freeman BA. Proc Natl Acad Sci USA 1990; 87:1620–1624.

9. Moncada S, Palmer RMJ, Higgs AE. Pharmacol Rev 1991;43:109–142.
10. White KA, Marletta MA. Biochemistry 1992;31:6627–6631.
11. Stuehr DJ, Ikeda-Saito M. J Biol Chem 1992;267:20547–20550.
12. McMillan K, Bredt DS, Hirsch DJ, Snyder SH, Clark JE, Masters BSS. Proc Natl Acad Sci USA 1992;89:11141–11145.
13. Klatt P, Schmidt K, Mayer B. Biochem J 1992;288:15–17.
14. Gruetter CA, Barry BK, McNamara DB, Gruetter DY, Kadowitz PJ, Ignarro LJ. J Cyclic Nucl Res 1979;5:211–224.

Molecular biology of nitric oxide synthase and structure of endothelial nitric oxide synthase gene

Yoshiki Yui[1], Miyahara Kaoru[2], Kazuhiro Sase[1], Takeshi Kawamoto[2], Katsumi Toda[2], Yoshinori Doi[3], Shohei Ogoshi[4], Ryuichi Hattori[1], Takeshi Aoyama[1], Yasutake Yamamoto[5], Kouzou Hashimoto[5], Rie-XiA Yang[2], Chuichi Kawai[6], Shigetake Sasayama[1] and Yutaka Shizuta[2]

[1]Third Division, Department of Internal Medicine, Faculty of Medicine, Kyoto University, Kyoto 606; [2]Department of Medical Chemistry, [3]Geriatric Medicine, [4]Surgery, [5]Internal Medicine, Kochi Medical School, Nankoku, Kochi 783; and [6]Kyoto University, Kyoto 606, Japan

Summary

Nitric oxide (NO), which accounts for the biological activity of endothelium-derived relaxing factor, is now considered to be an important molecule found throughout the various tissues and organs. NO is synthesized from L-arginine by nitric oxide synthase (NOS). So far, three NOS genes (Type I/neuronal NOS, Type II/inducible NOS, and Type III/endothelial NOS) have been identified and characterized.

We have cloned human endothelial NOS gene consisting of 26 exons in a total size of 23 kb. It has a "TATA-less" promoter which contains putative consensus sequences for activating protein-1 and -2, estrogen responsive element, and shear stress responsive element. There are several highly repetitive regions which could provide potential sites for DNA polymorphism.

Functions of NO

Nitric oxide (NO), the inorganic free-radical gas, is responsible for endothelium-derived relaxing factor activity. It is now considered to be an important molecule formed not only in the endothelial cells but also in many other cells having different functions [1–5]. Endothelial NO serves to control vascular tone and to maintain a nonthrombogenic surface through vasodilation and inhibition of platelet aggregation. Some types of neurons use NO as a neurotransmitter. NO derived from leukocytes and macrophages is considered to play a key role in cytotoxicity against microorganisms and tumor cells.

NO is produced from L-arginine by nitric oxide synthase (NOS; EC1.14.23) with the concomitant production of citrulline. This reaction requires NADPH, tetrahydrobiopterin, FMN, and FAD. It is inhibited by arginine analogues like nitro-arginine and methyl-arginine.

24

Type III (Endothelial NOS)

Isoform type III, endothelial NOS, has been cloned from bovine [25–27] and human [28,29] endothelial cells. It is a constitutive, calcium/calmodulin-regulated enzyme like type I enzyme. However, this isoform is smaller than type I NOS. Its open reading frame is 3.6 kb encoding 135 kDa protein.

In the endothelium, some receptor agonists and shear stress can stimulate NO synthesis. This NO diffuses into smooth muscle cells and causes vasodilation. NO also diffuses towards the lumen and aggregation [30].

Endothelial NOS is present in particulate form, but does not have any membrane-binding region. This isoform is bound to the membrane by the addition of myristate, a saturated fatty acid [31]. This binding has been confirmed by molecular analysis with site-directed mutagenesis [32,33]. Interestingly, endothelial enzyme can be translocated to the cytosolic fraction associated with phosphorylation of the enzyme following exposure to bradykinin [34]. These post-translational modification may be important in regulating the location of this enzyme.

Endothelial-NOS cloning

To elucidate the role of endothelial NOS gene in physiological and pathophysiological states, we recently isolated and characterized the human endothelial NOS gene (Miyahara K. et al., manuscript in prep.).

Recombinant cosmid clones were isolated by colony hybridization from the human genomic DNA library. Out of the 50 positive clones isolated, we selected the 36 kb-sized clone which contained the full length endothelial NOS gene (Fig. 2). Restriction fragments of cosmid DNA were subcloned into pUC118 vector and sequenced by the chemical dideoxy-termination method with the 373A automated

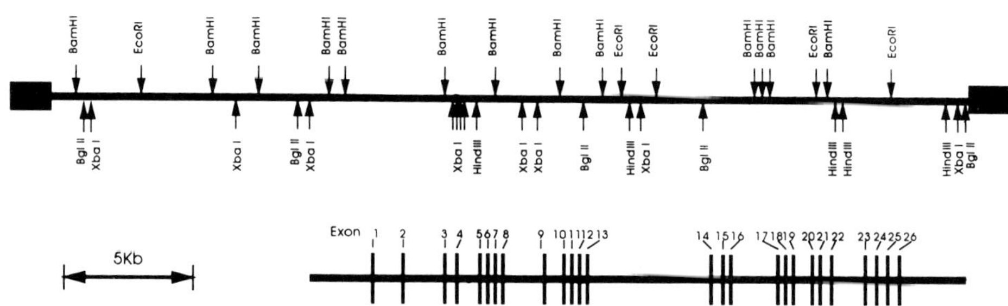

Fig. 2. Structure of human endothelial NOS gene. Restriction map of isolated cosmid clone (upper panel), and exon-intron alignment of NOS gene (lower panel). Exons are numbered and displayed by closed boxes.

sequencer (Applied Biosystems Inc.). We sequenced the full length of this gene including all introns. Southern blot analysis of digested human DNA showed that there is only one copy of endothelial NOS gene in the haploid human genome (data not shown).

Endothelial NOS gene consists of 26 exons in a total length of 23 kb (Fig. 2). All exon/intron junctions obey the GT-AG splice rule [35]. Interestingly, this gene has several highly repetitive regions. In introns 2 and 8 there are minisatellite repeats. Intron 4 contains a long direct repeat, and there is a highly polymorphic (CA)n repeat in intron 13.

In general, those repetitive elements could provide potential sites for DNA polymorphism [36]. Recently, some investigators have reported a relationship between DNA polymorphism and cardiovascular disease [37,38]. An angiotensino-gen gene has been found by an analysis of DNA polymorphism to be one of the genetic factors related to essential hypertension [39]. DNA polymorphism in the angiotensin converting enzyme gene provides new insights into the risk factors of myocardial infarction [40,41]. It might be of interest to examine the relationship between cardiovascular disorders, such as essential hypertension, atherosclerosis, and myocardial infarction, and the polymorphism of the endothelial NOS gene.

Analysis of the 5′-flanking region (Fig. 3)

There is no consensus TATA-box element located upstream of the transcription start site. However, there are several consensus sequences for the binding of Sp-1, which is related to the transcription of TATA-less genes [42]. There are many consensus sequences for activator protein-1, -2, and nuclear factor-kappa B near the transcription start site. Interestingly, the consensus sequence for the shear stress responsive element [43] is at about 1000 bp upstream. These putative gene

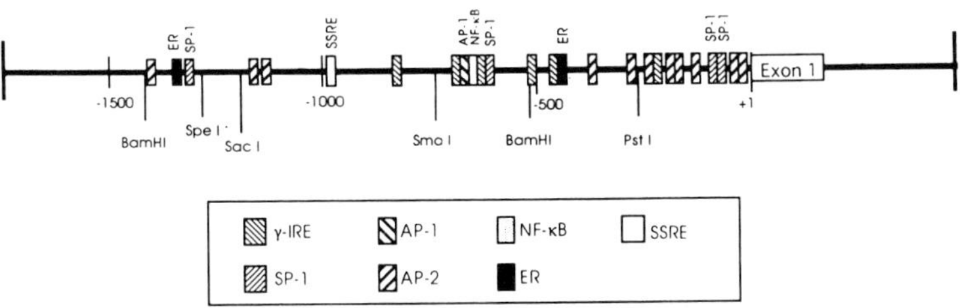

Fig. 3. Transcription factor binding consensus sequences in the 5′-flanking region of the human endothelial NOS gene: AP-1, activator protein 1; AP-2, activator protein 2; γIRE, gamma-interferon response element; NF-κB, nuclear factor kappa B; SSRE, shear stress response element; ER, estrogen response element (half).

regulatory elements suggest that transcription of endothelial NOS may be regulated by various cytokines as well as by shear stress.

In addition, two estrogen responsive half palindromic motifs [44] have been found. This is interesting, because estrogens have been thought to account for the lower incidence of atherosclerosis and coronary artery disease in pre-menopausal women than in men or postmenopausal women [45].

Our current observations may provide direction for future studies of endothelial NOS expression in physiological and pathophysiological conditions.

Chromosomal location of NOSs

Three distinct NOS genes have been identified so far.

Isoform type I, neuronal NOS, is located on chromosome 12 [46]. Type II, inducible NOS gene, has recently been isolated by another group [47] and mapped on chromosome 17. Type III, endothelial NOS is located on chromosome 7 [46].

It is uncertain whether or not other NOS isoforms are present in addition to these three. Further studies will be needed to clarify this point.

Acknowledgements

We express our appreciation to Dr. Alice S. Cary for help with preparation of the manuscript.

References

1. Furchgott RF, Vanhoutte PM. FASEB J 1989;3:2007–2018.
2. Moncada S, Palmer RM, Higgs EA. Pharmacol Rev 1991;43:109–142.
3. Nathan C. FASEB J 1992;6:3051–3064.
4. Lowenstein CJ, Snyder SH. Cell 1992;70:705-707.
5. Foestermann U, Pollock JS, Nakane M. Trends Cardiovasc Med 1993;3:104–110.
6. Bredt DS, Hwang PM, Glatt CE, Lowenstein C, Reed RR, Snyder SH. Nature 1991;351:714–718.
7. Nakane M, Schmidt HHW, Pollock JS, Foestermann U, Murad F. FEBS Lett 1993;316:175–180.
8. Garthwaite J. Trends Neurosci 1991;14:60–67.
9. Hiki K, Hattori R, Kawai C, Yui Y. J Biochem (Tokyo) 1992;111:556–558.
10. Yui Y, Hattori R, Kosuga K, Eizawa H, Hiki K, Kawai C. J Biol Chem 1991;266:12544–12547.
11. Yui Y, Hattori R, Kosuga K, Eizawa H, Hiki K, Ohkawa S, Ohnishi K, Terao S, Kawai C. J Biol Chem 1991;266:3369–3371.
12. Evans T, Carpenter A, Cohen J. Proc Natl Acad Sci USA 1992;89:5361–5365.
13. Busse R, Mulsch A. FEBS Lett. 1990;275:87–90.
14. Nathan CF, Hibbs JB. Curr Opin Immunol 1991;3:65–70.
15. Ochoa JB, Udekwu AO, Billiar TR, Curran RD, Cerra FB, Simmons RL, Peitzman AB. Ann Surg 1991;214:621–626.
16. Xie Q, Cho HJ, Calaycay J, Mumford RA, Swiderek KM, Lee TD, Ding A, Troso T, Nathan C. Science 1992;256:225–228.
17. Lyons CR, Orloff GJ, Cunningham JM. J Biol Chem 1992;267:6370–6374.

18. Lowenstein CJ, Glatt GS, Bredt DS, Snyder SH. Proc Natl Acad Sci USA 1992;89:6711–6715.

19. Nunokawa Y, Ishida N, Tanaka S. Biochem Biophys Res Comm 1993;191:89–94.

20. Wood ER, Berger H Jr., Sherman PA, Lapetina EG. Biochem Biophys Res Comm 1993;191: 767–774.

21. Geller DA, Lowenstein CJ, Shapiro RA, Nussler AK, DiSilvio M, Wang SC, Nakayama DK, Simmons RL, Snyder SH, Billiar TR. Proc Natl Acad Sci USA 1993;90:3491–3495.

22. Cho HJ, Xie QW, Calaycay J, Mumford RA, Swiderek KM, Lee TD, Nathan C. J Exp Med 1992; 176:599–604.

23. Xie Q, Whisnant R, Nathan C. J Exp Med 1993;177:1779–1784.

24. Lowenstein CJ, Alley EW, Raval P, Snowman AM, Snyder SH, Russell SW, Murphy WJ. Proc Natl Acad Sci USA 1993;90:9730–9734.

25. Lamas S, Marsden PA, Li GK, Tempst P, Michel T. Proc Natl Acad Sci USA 1992;89:6348–6352.

26. Sessa WC, Harrison JK, Barber CM, Zeng D, Durieux ME, D'Angelo DD, Lynch KR, Peach MJ. J Biol Chem 1992;267:15274–15276.

27. Nishida K, Harrison DG, Navas JP, Fisher AA, Dockery SP, Uematsu M, Nerem RM, Alexander RW, Murphy TJ. J Clin Invest 1992;90:2092–2096.

28. Janssens SP, Shimouchi A, Quertermous T, Bloch DB, Bloch KD. J Biol Chem 1992;267: 14519–14522.

29. Marsden PA, Schappert KT, Chen HS, Flowers M, Sundell CL, Wilcox JN, Lamas S, Michel T. FEBS Lett 1992;307:287–293.

30. Radomski MW, Plamer RMJ, Moncada S. Lancet 1987;2:1057–1058.

31. Pollock JS, Klinghofer V, Foestermann U, Murad F. FEBS Lett 1992;309:402–404.

32. Busconi L, Michel T. J Biol Chem 1993;268:8410–8413.

33. Sessa WC, Barber CM, Lynch KR. Circ Res 1993;72:921–924.

34. Michel T, Gordon KL, Busconi L. Proc Natl Acad Sci USA 1993;90:6252–6256.

35. Mount SM. Nucleic Acids Res 1982;10:459–472.

36. Keating M. Circulation 1992;85:1973–1986.

37. Tanigawa G, Jarcho JA, Kass S, Solomon SD, Vosberg HP, Seidman JG, Seidman CE. Cell 1990; 62:991–998.

38. Lee B, Godfrey M, Vitale E, Hori H, Mattei MG, Sarfarazi M, Tsipouras P, Ramirez F, Hollister DW. Nature 1991;352:330–334.

39. Jeunemaitre X, Soubrier F, Kotelevtsev YV, Lifton RP, Williams CS, Charru A, Hunt SC, Hopkins PN, Williams RR, Lalouel JM, Corvol P. Cell 1992;71:169–180.

40. Rigat B, Hubert C, Alhenc-Gelas F, Cambien F, Corvol P, Soubrier F. J Clin Invest 1990;86: 1343–1346.

41. Cambien F, Poirier O, Lecerf L, Evans A, Cambou JP, Arveiler D, Luc G, Bard JM, Bara L, Ricard S, Tiret L, Amouyel P, Alhenc-Gelas F, Soubrier F. Nature 1992;359:641–644.

42. Faber PW, van Rooij HC, Schipper HJ, Brinkmann AO, Trapman J. The role of Sp1. J Biol Chem 1993;268:9296–9301.

43. Resnick N, Dewey CF, Atkinson W, Collins T, Gimbrone MA Jr. Proc Natl Acad Sci USA 1993; 90:4591–4595.

44. Peale FV Jr, Ludwig LB, Zain S, Hilf R, Bambara RA. Proc Natl Acad Sci USA 1988;85: 1038–1042.

45. Stampfer MJ, Colditz GA, Willett WC, Manson JE, Rosner B, Speizer FE, Hennekens CH. N Engl J Med 1991;325:756–762.

46. Marsden PA, Heng HHQ, Scherer SW, Stewart RJ, Hall AV, Shi XM, Tsui LC, Schappert KT. J Biol Chem 1993;268:17478–17488.

47. Chartrain NA, Geller D, Koty P, Sitrin NF, Nussler AK, Hoffman EP, Billiar EP, Hutchinson NI, Mudgett JS. Molecular cloning, structure, and chromosomal localization of the human inducible nitric oxide synthase gene. J Biol Chem 1994;(in press).

Endothelium-Derived Factors and Vascular Functions. T. Masaki, ed.

The nitric oxide-cyclic GMP signal transduction system

Ferid Murad

Molecular Geriatrics Corporation, Lake Bluff, IL, USA

Summary

The formation of EDRF/NO from endogenous arginine is a ubiquitous signal transduction pathway. This enzymatic pathway is catalyzed by a family of isoforms of NO synthase. To date four major classes of isoforms have been characterized. The enzymes may be soluble or particulate, may be constitutive or inducible with cytokines and endotoxin and some isoforms are calcium/calmodulin dependent. Many of these isoforms have been purified and characterized in our laboratory and other laboratories. Polyclonal and monoclonal antibodies have permitted us to examine immunohistochemically cellular and tissue distribution of these isoforms. The Type I isoform is found in brain and other neuronal tissues including nonadrenergic noncholinergic neurons. This isoform is also found in non-neuronal cells such as lung and gastrointestinal epithelial cells, endometrium and pancreatic islet cells with immunohistochemical studies. This constitutive enzyme is NADPH, tetrahydrobiopterin, and calcium/calmodulin dependent. The Type III isoform is also constitutive, NADPH, tetrahydrobiopterin and calcium/calmodulin dependent. This isoform is responsible for endothelial-dependent relaxation in numerous vascular preparations. Type II (soluble) and Type IV (particulate) isoforms are inducible with endotoxin and some cytokines. These isoforms, originally found in macrophages, are ubiquitous in smooth muscle and other tissues after cytokine induction. Recent studies suggest that the inducible isoforms are also regulated by calmodulin. The products of these reactions are NO and citrulline. Nitric oxide activates soluble guanylate cyclase in numerous tissues, including vascular, airway and gastrointestinal smooth muscle, to increase the synthesis of the intracellular messenger cyclic GMP. The formed nitric oxide may act within the cell in which it is generated or in adjacent cells to increase cyclic GMP levels. Thus, NO can function as either an intracellular or an intercellular substance to participate in signal transduction through increased cyclic GMP synthesis. In most smooth muscle preparations, increases in cyclic GMP lead to the activation of cyclic GMP-dependent protein kinase, altered phosphorylation of many endogenous proteins, decreased cytosolic calcium, decreased phosphorylation of myosin light chain and relaxation. Some of the effects of nitric oxide and/or cyclic GMP may be mediated through other mechanisms.

Introduction

This brief review will describe some of the observations that have led to our present understanding of the nitric oxide-cyclic GMP signal transduction system. Readers are also referred to some other reviews for additional information and references [1–6]. While cyclic GMP has been viewed as a promising and potentially important second messenger for many years, our work, and that of others in the past 15 years, with nitric oxide and/or cyclic GMP has elevated the significance of these molecules in signal transduction within individual cells and between cells.

Nitric oxide has emerged as a rather simple and unique molecule that can serve many diverse functions including an intracellular second messenger and an intercellular messenger (paracrine substance, autacoid or hormone) to regulate neighboring cells. Signals from circulating nitric oxide could perhaps also be transmitted to distant cellular targets from the cell of origin for nitric oxide. This could occur if complexes or carrier states for nitric oxide existed that could release or regenerate nitric oxide at its distant target, in some ways analagous to hormone ligands bound to carrier proteins and molecules that behave as inactive complexes until the hormone is released or dissociated at a target. The ubiquity and reactivity of nitric oxide with thiols, proteins, sugars, metals, heme proteins, etc., will permit us to predict with some degree of certainty, that nitric oxide complexes and adducts will, undoubtedly, be present in various extracellular fluids. The question is, do any of these complexes serve a humoral function. We suspect that they will, considering the low concentrations of nitric oxide required to activate guanylyl cyclase and elevate cyclic GMP levels in tissues (see below).

Nitric oxide is also formed by various neuronal cells, both centrally and peripherally, and therefore, functions as a neurotransmitter. Such a diverse role for an individual intracellular and intercellular messenger has not been previously described.

Effects of nitric oxide on cyclic GMP synthesis and smooth muscle relaxation

From the earliest work in our laboratory and, subsequently, other laboratories, it has been known that cyclic GMP induces the relaxation of numerous smooth muscle preparations including vascular, airway and intestinal smooth muscle. Smooth muscle relaxation was one of the first physiological functions clearly related to cyclic GMP synthesis. The proposed functions of cyclic GMP have expanded considerably since, as briefly discussed below.

The increase in cyclic GMP in smooth muscle preparations is associated with cyclic GMP-dependent protein kinase activation and altered phosphorylation of numerous endogenous smooth muscle proteins. The functions of most of these phosphoproteins have not been determined. Cyclic GMP can also decrease the activity of phospholipase C in vascular preparations and smooth muscle cell cultures as determined by decreased formation of inositol phosphates and this effect also appears mediated by cyclic GMP-dependent protein kinase [7,8]. These effects probably lower cytosolic free calcium concentrations resulting in the decreased phosphorylation of myosin light chain and relaxation. Other interesting and hypothesized target sites for cyclic GMP regulated effects in smooth muscle could perhaps be membrane transport of calcium and/or other cations and protein phosphotase regulation. Obviously, many additional studies are warranted.

Other functions of cyclic GMP have included phototransduction in the retina [9], enterotoxin-induced intestinal secretion [10], inhibition of platelet aggregation and a variety of less clearly defined effects.

An appreciation of the smooth muscle relaxant effects of cyclic GMP came from our earlier studies with azide, nitrite, hydroxylamine, and numerous "nitrovasodilators" which increased cyclic GMP synthesis and relaxed a variety of smooth muscle preparations [11–14]. These agents share a common feature in that they either liberate nitric oxide under the appropriate redox conditions or can be enzymatically converted to this reactive free radical which activates soluble guanylyl cyclase. These studies have since been confirmed by numerous laboratories working with a variety of tissue preparations. We coined the term "nitrovasodilators" for this broad class of nitric oxide forming agents even though many of these agents do not posses a nitro or nitroso functionality and require oxidation and enzymatic conversion.

It is rather uniformly believed that only the soluble isoform of guanylyl cyclase can be activated by nitrovasodilators. The soluble isoform is a heterodimer [15] and each dimer appears to possess a catalytic domain from cloning experiments [16,17]. While the enzyme possesses a heme moiety which is thought to participate in the activation mechanism [18–20], the actual mechanism of activation has not been definitively determined since this requires large quantities of active protein for the appropriate physiochemical and spectral studies.

While soluble guanylyl cyclase is clearly activated by the nitric oxide generating nitrovasodilator agents, some of the particulate isoforms of guanylyl cyclase can also be activated. These include enzyme preparations from retina, intestinal epithelium and liver [6,10,21]. The particulate isoforms of guanylyl cyclase are clearly different structurally from the soluble enzyme [2–4,10,22]. However, some common regulatory features may be shared. Perhaps some of the particulate isoforms also contain a heme moiety, a transition metal or critical sulfhydryl groups that interact with NO to cause conformational changes in the enzyme and activation. Clearly, numerous experiments are yet to be conducted.

Another interesting feature of soluble guanylyl cyclase is that while each subunit appears to possess a catalytic domain, both subunits are required for catalytic activity and nitric oxide activation [17,23]. Our mutagenesis studies with soluble guanylyl cyclase are inconclusive to clarify the complexities of the subunit interactions.

After finding that nitrovasodilators mediated their effects through nitric oxide production and increased cyclic GMP synthesis, we proposed that nitric oxide could be a natural endogenous agent whose formation from some endogenous precursor(s) could be hormonally regulated in order to explain hormonal regulation of cyclic GMP synthesis [24,25]. Since nanamolar concentrations of NO are capable of activating guanylyl cyclase and increasing cyclic GMP, and all of the assays for NO and nitrite at that time were several orders of magnitude less sensitive, it was not possible to actively test this hypothesis with definitive experiments. It took another decade and work from several independent directions before this hypothesis was proven.

32

Effects of EDRF an "endogenous nitrovasodilator"

Furchgott and his associates found that some vasorelaxants required the presence of the endothelium to cause the release of an endothelium derived relaxant factor (EDRF) and vascular relaxation [26,27]. Subsequently, we found that the effects of EDRF were also mediated through increased cyclic GMP synthesis in the smooth muscle cells of blood vessels [28–31]. Furthermore, endothelium-dependent vasodilators increased cyclic GMP-dependent protein kinase activity and altered the phosphorylation of the same proteins as identified on 2D-gels as did nitrovasodilators including the decrease in myosin light chain phosphorylation. Thus, the pharmacological and biochemical effects of endothelium-dependent vasodilators and nitrovasodilators were virtually identical and we came to view EDRF as an "endogenous nitrovasodilator" [1]. Shortly thereafter, Furchgott [32] and Ignarro [33] proposed that EDRF could be nitric oxide. While nitric oxide synthase can indeed convert arginine to citrulline and nitric oxide as discussed below, nitric oxide may form a complex with some endogenous material(s) before or after it is liberated from cells. Recent studies suggest that this nitric oxide complex may contain a thiol [34]. Presumably, this complex behaves as EDRF and subsequently, liberates nitric oxide which activates guanylyl cyclase. This hypothesis is much more appealing to us, particularly if there were specific and/or selective transport and delivery mechanisms for this EDRF complex. The marked reactivity of the nitric oxide free radical with numerous substances [35] seems to preclude NO itself being EDRF. Nitric oxide at the nanamolar concentrations that are required for this signal transduction process would probably fail to reach and activate soluble guanylyl cyclase in a neighboring targeted cell before colliding with other molecules to form an inactive complex. The chemistry of nitric oxide formation and reactivity is quite complex and this has become increasingly important in recent studies. We demonstrated a number of years ago that the dose response curve for NO activation of purified guanylyl cyclase is markedly shifted to the right in the presence of albumin, thiols, sugars and other substances [35]. Unfortunately, the low concentrations of EDRF in tissues and its reactivity and short half life will require some novel and challenging chemistry to identify the specific structure of EDRF which undoubtedly contains NO. It is not anticipated that the precise structure of EDRF will be definitively proven in the near future.

Effects of L-arginine on cyclic GMP and nitrite formation

DeGuchi's laboratory [36] reported that partially purified preparations of guanylyl cyclase from neuroblastoma cells could be activated by an endogenous factor in brain extracts that proved to be L-arginine. The stimulatory effects of L-arginine were lost with highly purified guanylyl cyclase. The significance of these observations was not fully appreciated at the time. However, the authors concluded that the mechanism of activation with arginine must be similar to that of nitroso compounds ("nitrovasodilators") since their stimulatory effects were not addictive and were blocked with similar agents such as methylhydroxylamine and hemo-globin.

Quite independently, Hibbs et al. [37] found that macrophages could produce nitric oxide and nitrite which was associated with their cytotoxic properties. This laboratory also reported that some arginine analogues could prevent NO/nitrite formation. This observation was rapidly recognized by many laboratories that proceeded to show that arginine was converted to citrulline and nitric oxide in numerous cell-free and intact cell systems including macrophages, leukocytes, endothelial cells, brain, etc., This enzyme has been referred to as NO synthase, EDRF synthase, or guanylyl cyclase activating factor (GAF) synthase [38–40]. As discussed below, most cells and tissues, but not all, possess this synthetic pathway.

Isoforms of nitric oxide synthase

Today it is known that there are many isoforms of this enzyme which are found in numerous tissues (Table 1) [41–44]. The enzyme may be soluble or particulate and it may be constitutive or inducible with endotoxin and some cytokines. The constitutive isoforms are regulated by Ca^{2+} calmodulin. All of the isoforms require NADPH, tetrahydrobiopterin, FAD and FMN (see Table 1).

Several isoforms have been purified to homogeneity and antibodies have been generated. The soluble brain enzyme (Type I), the soluble inducible macrophage

Table 1. Isoforms of NO synthase. All isoenzymes use L-arginine as a substrate and all are inhibited by N^G-methyl-L-arginine and N^G-nitro-L-arginine. The Type Ia, II, and III have been cloned by several laboratories, show about 50 to 60% homology and obviously represent separate gene products. Some isoforms may also represent post-translational modifications. It is also recognized that this classification is probably incomplete as additional isoforms are expected from future purification, cloning and post-translational modification experiments.

Type	Cosubstrates Co-factors	Regulated by	M_r	Present in
I a (soluble)	NADPH, BH_4, FAD/FMN*	Ca^{2+}/calmodulin	155 kDa	Brain, cerebellum N1E-115 neuroblastoma cells
I b (soluble)	NADPH	Ca^{2+}/calmodulin	135 kDa	Endothelial cells
I c (soluble)	NADPH, BH_4, FAD	Ca^{2+} (*not* calmodulin)	150 kDa	Neutrophils
II (soluble)	NADPH, BH_4, FAD/FMN	Unknown (induced by endotoxin/cytokines)	125 kDa	Macrophages and other cells
III (particulate)	NADPH, BH_4, FAD/FMN	Ca^{2+}/calmodulin	135 kDa	Endothelial cells
IV (particulate)	NADPH	Unknown (induced by endotoxin/cytokines)	?	Macrophages and other cells

34

enzyme (Type II), and the particulate endothelial enzyme (Type III) have been cloned by several different laboratories and cloning activity continues with the other isoforms in a number of laboratories. In addition to the regulation of catalytic activity by the cofactors listed above and Ca^{2+}/calmodulin, there is some evidence that phosphorylation of the enzyme by various kinases can increase and/or decrease activity [45]. The regulatory significance, if any, with phosphorylation of the enzyme to control catalytic activity is not presently known. We have found that the particulate isoform from endothelial cells (Type III) is myristoylated which probably explains its association with membranes [46]. Post-translational modifications of other isoforms will probably also be found. The ultimate numbers of individual gene products, possible post-translational modifications and their functional significance on subcellular localization and catalytic capacity are presently unknown and the topics of active investigation.

Immunohistochemical localization of the Type I and Type III isoforms with selective polyclonal and monoclonal antibodies that we and others have generated demonstrates a rather ubiquitous distribution of these isoforms in numerous tissues [47,48]. Our polyclonal antibodies to the brain Type I enzyme are quite selective and do not recognize the Type III enzyme from endothelial cells. Our monoclonal antibodies to endothelial Type III enzyme may be selective for the Type III enzyme or may crossreact with common epitopes shared by the Type I and Type III enzyme. Many laboratories have since developed antibodies to either purified proteins or synthetic peptide fragments.

We, and others, have also found that nonadrenergic-noncholinergic (NANC) innervated tissues such as bovine retractor penis and rat anococcygeus possess a brain Type I NO synthase activity which we partially purified and characterized [49,50]. Furthermore, nerve fibers in these tissues as well as gastrointestinal tissues demonstrate immunohistochemical staining with antibody to the Type I brain NO synthase. Thus, nitric oxide is probably one of the neurotransmitters of NANC neurons and the presence of NANC or "nitrinergic neurons" is quite ubiquitous in the peripheral as well as the central nervous system.

Summary

From our work and the work of others, it is now quite apparent that the nitric oxide-cyclic GMP system can function as an intracellular or an intercellular signal transduction system [2–5,51,52]. If a specific cell possesses both nitric oxide synthase and an isoform of guanylyl cyclase that is activatable with NO, then cyclic GMP levels in that cell can be regulated by agents that alter nitric oxide synthase activity and NO formation (Figure 1). NO, or a complex of NO, which is liberated from the producing or donor cell, can also activate guanylyl cyclase in a neighboring or perhaps a distant cell to increase cyclic GMP synthesis. In the latter scenario, NO or its carrier complex behaves as a paracrine substance, autacoid or hormone. Interestingly, the liberated extracellular NO can also feed back and increase cyclic GMP synthesis in the cell of origin. This is best demonstrated by the inhibitory effects of hemoglobin on agonist-induced cyclic GMP accumulation in homogenous cell culture systems where the hormone or agonist effects on cyclic GMP are mediated by NO. Presumably, hemoglobin would not be permeable and could only trap or scavenge extracellular NO to account for its ability to decrease hormonally-induced cyclic GMP increases in homogenous cell populations. There is no direct evidence that NO can act as an endocrine substance to increase cyclic GMP synthesis in a distant target cell population. However, complexes or carrier states of NO that could liberate NO at a distant site, could most certainly be viewed as endocrinologic agents (hormones or antacids). We suspect that appropriately designed experiments in the future will also support this role for NO as an endocrinologic agent that can also function at a distance similar to classical hormones.

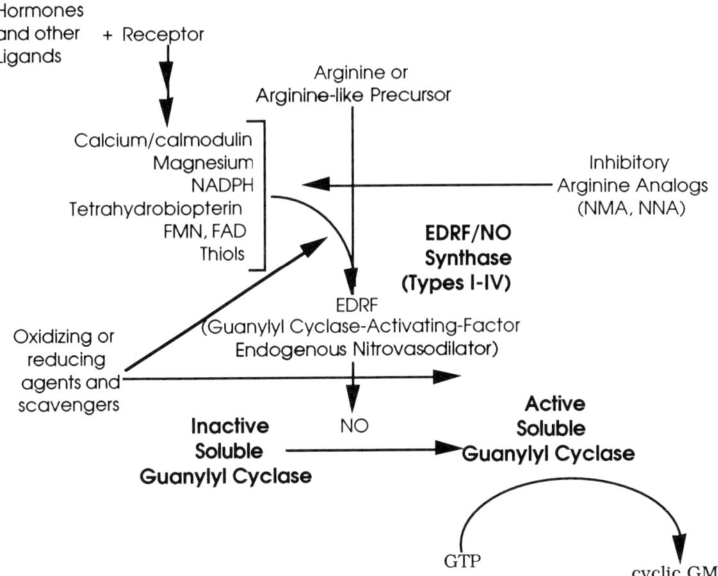

Fig. 1. The nitric oxide/cyclic GMP signal transduction pathway. The nitric oxide formed from the oxidation of the guanidino nitrogen of arginine can act as an intracellular or intercellular messenger to regulate cyclic GMP synthesis. Some effects of nitric oxide may be mediated independently of cyclic GMP synthesis.

Indeed, we believe that nitric oxide should be added to the list of agents that can function as a neurotransmitter, paracrine substance and autacoid or hormone. It can also be viewed as an intracellular, as well as intercellular, messenger. To date, no substance has played such a diverse role in intracellular and intercellular signal transduction. Thus, NO appears to be a unique and simple molecule with diverse functions in signal transduction.

36

Although many laboratories and investigators were skeptical of the possible importance and roles for cyclic GMP and nitric oxide some years ago when we initially described their interrelationships and proposed some of their roles in signal transduction, this field has since attracted considerable attention.

Acknowledgments

Numerous trainees and collaborators have participated in our studies with the nitric oxide-cyclic GMP system in a wide variety of tissues and systems in the past 15 years. The discussions, debates, and progress have been rewarding to all of us. Many of the studies summarized here were supported with grants from the NIH and many other agencies.

References

1. Murad F. J Clin Invest 1986;78:1-5.
2. Murad F, Leitman D, Waldman S, Chang CH, Hirata J, Kohse K. Proc Cold Spring Harbor Symposium on Quantitative Biology, Signal Transduction 1988;53:1005-1009.
3. Murad F. In: Gehring Y, Helmreich E, Schultz G (eds.) Molecular Mechanisms of Hormone Action. Heidelberg: Springer, 1989;186-194.
4. Murad F. In: Catravas JD, Gillis CM, Ryan US (eds.) Vascular Endothelium Plenum Publishers, 1989;157-164.
5. Murad F, Ishii K, Forstermann U, Gorsky L, Kerwin J, Pollock J, Heller M. Adv Cyclic Nucl Res 1990;24:441-448.
6. Waldman SA, Lewicki JA, Brandwein HJ, Murad F. J Cyclic Nuc Res 1982;8:359-370.
7. Rapoport RM. Circ Res 1986;58:407-410.
8. Hirata M, Kohse K, Chang CH, Ikebe T, Murad F. J Biol Chem 1990;265:1268-1273.
9. Stryer L. Annu Rev Neuroscience 1986;9:87-119.
10. Waldman SA, Murad F. Pharm Rev 1987;39:163-196.
11. Katsuki S, Arnold WP, Murad F. J Cyclic Nucl Res 1977;3:239-247.
12. Katsuki S, Arnold W, Mittal CK, Murad F. J Cyclic Nucl Res 1977;3:23-35.
13. Kimura H, Mittal CK, Murad F. J Biol Chem 1975;250:8016-8022.
14. Kimura H, Mittal CK, Murad F. Nature 1975;257:700-702.
15. Kamisaki Y, Saheki S, Nakane M, Palmieri J, Kuno T, Chang B, Waldman SA, Murad F. J Biol Chem 1986;261:7236-7241.
16. Nakane M, Saheki S, Kuno T, Ishii K, DeGuchi T, Murad F. Biochem Biophys Res Commun 1988;157:1139-1147.
17. Nakane M, Arai K, Saheki S, Kuno T, Buechler W, Murad F. J Biol Chem 1990;265:16841-16845.
18. Gerzer R, Bohme E, Hoffman F, Schultz G. FEBS Lett 1981;132:71-74.
19. Ignarro LJ, Adams JB, Horwitz PM, Wood KS. J Biol Chem 1986;261:4997-5002.
20. Lewicki JA, Brandwein HJ, Mittal CK, Arnold WP, Murad F. J Cyclic Nucl Res 1982;8:17-25.
21. Horio Y, Murad F. Biochim Biophys Acta 1991;1122:81-88.
22. Garbers DL. Can J Physiol Pharmacol 1991;69:1618-1621.
23. Saheki S, Kuno T, Takeuchi N, Murad F. Biochim Biophys Acta 1990;1051:306-309.
24. Murad F, Mittal CK, Arnold WP, Braughler JM. In: Stocklet JD (ed.) Advances in Pharmacology and Therapeutics, Vol e Ions, Cyclic Nucleotides, Cholinergy, New York: Pergamon Press, 1978; 2123-2132.
25. Murad F, Mittal CK, Arnold WP, Katsuki S, Kimura H. Adv Cyclic Nucl Res 1978;9:145-158.

26. Furchgott RF, Sawadski, JV. Nature 1980;288:373-376.
27. Furchgott RF, Vanhoutte PM. FASEB J 1989;3:2007-2018.
28. Rapoport RM, Murad F. Circ Res 1983;52:352-357.
29. Rapoport RM, Murad F. J Cyclic Nucl and Protein Phosphor Res 1983;9:281-296.
30. Rapoport RM, Draznin MD, Murad F. Trans Assoc Amer Phys 1983;96:19-30.
31. Fiscus RR, Rapoport RM, Murad F. J Cyclic Nucl Res 1983;9:415-425.
32. Furchgott RF. In: Vanhoutte PM (ed.) Vasodilation: Vascular Smooth Muscle, Peptides, Autonomic Nerves and Endothelium. New York: Raven Press, 1988;401-414.
33. Ignarro LJ, Buga GM, Wood KS, Byrns RE, Chadhuri G. Proc Natl Acad Sci 1987;84:9265-9269.
34. Myers PR, Minor RL, Guerro R, Bates JN, Harrison DG. Nature 1990;345:161-163.
35. Braughler JJ, Mittal CK, Murad F. J Biol Chem 1979;254:12450-12454.
36. DeGuchi T, Yoshioka M. J Biol Chem 1982;257:10147-10151.
37. Hibbs JR, Taintor RR, Varrin Z. Science 1987;235:473-476.
38. Bredt DS, Snyder SH. Proc Natl Acad Sci USA 1990;85:682-685.
39. Schmidt HHHW, Pollock J, Nakane M, Gorsky L, Forstermann U, Heller M, Murad F. Proc Natl Acad Sci 1991;88:365-369.
40. Stuehr DJ, Cho HJ, Kwon NS, Wise MF, Nathan CF. Proc Natl Acad Sci USA 1991;88:7773-7777.
41. Forstermann U, Gorsky L, Pollock JS, Schmidt HHHW, Ishii K, Heller M, Murad F. Eur J Pharmacol 1990;183:1625-1626.
42. Forstermann U, Pollock J, Schmidt HHHW, Heller M, Murad F. Proc Natl Acad Sci 1991;88:1788-1792.
43. Forstermann U, Schmidt HHHW, Pollock JS, Sheng H, Mitchell JA, Warner TD, Nakane M, Murad F. Biochem Pharmacol 1991;41:1849-1857.
44. Palmer R, Ashton D, Moncada S. Nature 1988;333:664-665.
45. Nakane M, Mitchell JA, Forstermann U, Murad F. Biochem Biophys Res Commun 1991;180:1396-1402.
46. Pollock JS, Klinghofer V, Forstermann U, Murad F. FEBS Lett 1992;309:402-404.
47. Snyder SH, Bredt DS. Trends Pharmacol Sci 1991;12:125-130.
48. Schmidt H, Gagne G, Nakane M, Pollock J, Miller M, Murad F. J Histo Cytochem 1992;40:1439-1456.
49. Mitchell JA, Sheng H, Forstermann U, Murad F. Brit J Pharmacol 1991;104:289-291.
50. Sheng H, Nakane M, Schmidt HHHW, Mitchell J, Pollock J, Forstermann U, Murad F. Brit J Pharmacol 1992;106:768-773.
51. Ishii K, Gorsky L, Forstermann U, Murad F. J Applied Cardiology 1989;4:505-512.
52. Ishii K, Chang B, Kerwin JF, Wagenaar FL, Huang ZJ, Murad F. J PET 1991;256:38-43.

Endothelium-Derived Factors and Vascular Functions. T. Masaki, ed.

Role of nitric oxide (NO) derived from endothelium and vasodilator nerve for the regulation of cerebral arterial tone

Noboru Toda

Department of Pharmacology, Shiga University of Medical Sciences, Seta, Ohtsu 520-21, Japan

Summary

Nicotine, substance P and arginine vasopressin injected into the vertebral artery produce a basilar arterial dilatation in anesthetized dogs by a mediation of nitric oxide (NO). Studies on isolated dog cerebral arteries indicate that the nicotine-induced relaxation is endothelium-independent and is mediated by NO derived from vasodilator nerve that is thus called "nitroxidergic". Relaxations caused by substance P and vasopressin are mediated by endothelium-derived NO. Intracisternal injections of NO synthase inhibitors elicit basilar arterial constriction in anesthetized dogs, which is markedly reduced by hexamethonium. Under the experimental conditions used, NO released from the nitroxidergic nerve appears to participate in dilating cerebral arteries predominantly over NO released from the endothelium.

Introduction

Nitric oxide (NO) is now regarded as a most attractive messenger in regulating the vascular function. Endothelial cells and macrophages are first determined to synthesize and liberate NO, which is also recognized to be produced in central and peripheral nerves and smooth muscle cells. The purpose of this contribution is to compare actions of nicotine, substance P and arginine vasopressin on basilar arterial diameter in anesthetized dogs, to analyze the mechanism of action of these substances in isolated dog cerebral arteries, and to clarify the control by NO derived from the nerve and endothelium of basilar arterial tone in anesthetized dogs.

Effect of NO releasing substances on basilar arterial diameter in vivo

In anesthetized dogs, transfemoral vertebral angiography was performed with a digital subtraction angiography system. Nicotine or substance P was injected into the vertebral artery, and the basilar arterial diameter was angiographically measured before and 10 seconds after the injection.

Bolus injections of nicotine (10^{-4} M, 1 ml) clearly dilated the basilar artery [1]. The dilatation disappeared shortly. The same volume of saline did not alter the arterial diameter. During the period of the vasodilatation, no change was observed in the systemic blood pressure, heart rate and respiration. Summarized in Fig. 1 are

40

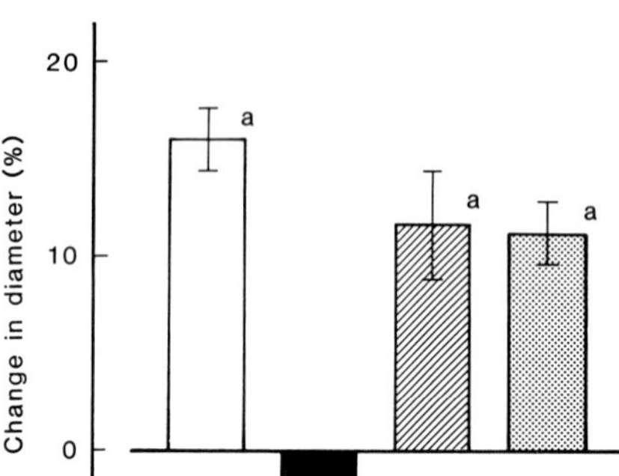

Fig. 1. Effects of nicotine (10^{-4} M, 1 ml) on basilar arterial diameter angiographically measured in anesthetized dogs before and after intracisternal treatment with L-NA, D-NA and L-arginine (L-arg). Reproduced from Toda et al., Am J Physiol 1993; 265: H103–H107, #1.

the quantitative data. The ordinate represents changes expressed as values relative to basilar arterial diameter before the nicotine injection. Nicotine dilated the artery by about 17%. The response was abolished or reversed to a constriction by treatment of the dogs with intracisternal N^{G}-nitro-L-arginine (L-NA), an NO synthase

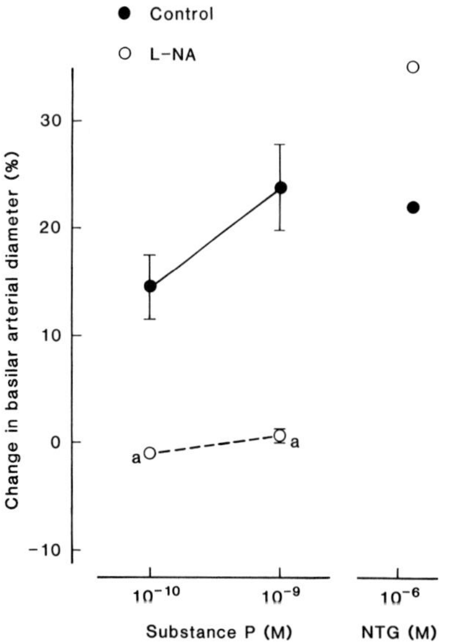

Fig. 2. Effects of substance P (10^{-10} and 10^{-9} M, 1 ml, $n = 4$) and nitroglycerin (NTG; 10^{-6} M, 1 ml, $n = 2$) on basilar arterial diameter in anesthetized dogs before and after treatment with L-NA (0.4 mM, 5 ml, intracisternal).

inhibitor. This inhibitory effect was reversed by the addition of L-arginine. On the other hand, N^G-nitro-D-arginine (D-NA) did not significantly alter the effect of nicotine. It is concluded that the nicotine-induced vasodilatation is mediated by NO synthesized from L-arginine by NO synthase.

In the same method, we determined the effect of substance P on basilar arteries. This peptide (10^{-10} and 10^{-9} M, 1 ml) intravertebrally injected produced a dose-related dilatation, 14 and 22%, respectively (Fig. 2). Intracisternal application of L-NA abolished the response to substance P. In the right panel, you can see the dilator effect of nitroglycerin in a dose sufficient to produce the effect similar to that caused by 10^{-9} M substance P. This effect was not attenuated by treatment with L-NA. NO synthesized from L-arginine via NO synthase appears to be involved in the vasodilatation associated with substance P but not the nitroglycerin induced dilatation.

Similar results were obtained by Suzuki et al. (2) in anesthetized dogs in response to arginine vasopressin. Intravertebral injections of the peptide dilated the basilar artery. Intracisternal treatment with L-monomethyl arginine, another NO synthase inhibitor, almost abolished the vasodilator effect of vasopressin. In contrast, the vasodilatation caused by vasoactive intestinal polypeptide (VIP) was not influenced by the NO synthase inhibitor. The effect of vasopressin, but not vasoactive intestinal polypeptide is expected to derive from NO synthesized by NO synthase.

These results indicate that the basilar arterial dilatation caused by nicotine, substance P and vasopressin in anesthetized dogs is associated with NO synthesized by NO synthase. However, in such an in vivo system, whether NO was derived from the endothelium, nerve or else could not be determined. Therefore, mechanisms of the vasodilator action of these substances were analyzed in isolated dog cerebral arteries.

Mechanism of action of nicotine, substance P and vasopressin

Nicotine

Helical strips of dog basilar and middle cerebral arteries with and without the endothelium were suspended in the bathing media for isometric tension recordings. Drugs were applied directly to the bathing media. If necessary, the strips were transmurally stimulated by electrical pulses via the platinum electrodes.

The addition of nicotine (10^{-4} M), NO (10^{-7} M) and nitroglycerin (10^{-8} M) moderately relaxed basilar arterial strips contracted with prostaglandin (PG) $F_{2\alpha}$. Treatment with L-NA reversed the nicotine-induced relaxation to a contraction. D-NA had no effect. The relaxation inhibited by L-NA was reversed by additional treatment with L-arginine but not by D-arginine [3]. On the other hand, relaxations caused by NO and nitroglycerin were not influenced by L-NA and L-arginine. Our accumulated data indicate that the response to nicotine is not inhibited by atropine, β-adrenoceptor antagonists, aminophylline and indomethacin but is abolished by hexamethonium, methylene blue, an inhibitor of soluble guanylate cyclase, and oxyhemoglobin [4–6]. Endothelium denudation did not change the response to nicotine [3]. L-NA almost abolished the relaxation in the strips with denuded endothelium, as in the endothelium-intact strips. It is evident that nicotine-induced relaxations are endothelium-independent, and actions of nitro-arginine and arginine are stereo-specific.

Transmural electrical stimulation at frequencies of 2, 5 and 20 Hz applied to dog cerebral arteries denuded of the endothelium produced relaxations which were not influenced by D-NA or D-monomethyl arginine but were abolished by their L-enantiomers [3,7]. L-arginine restored the response which was abolished by tetrodotoxin. It is concluded that relaxations evoked by stimulation of vasodilator nerves with electrical pulses as well as nicotine are mediated by NO produced via NO synthase.

The arterial relaxation induced by nicotine was attenuated by cadmium (10^{-5} and 5×10^{-5} M) in a dose-dependent manner, but was not significantly influenced by nicardipine, a dihydropyridine Ca^{2+} entry blocker, in a concentration sufficient to produce a suppression of the contractile response of dog cerebral arteries to $PGF_{2\alpha}$ and serotonin to a similar extent to that caused by 5×10^{-5} M cadmium (8). The NO-induced relaxation was not affected by these Ca^{2+} antagonists. Removal of external Ca^{2+} also suppressed the response to nicotine. The inhibition by cadmium of the response to nicotine appears to be associated with the interference with the Ca^{2+} influx through Ca^{2+} channels distinct from the channel of L-type.

Nicotine (10^{-4} M), nitric oxide (10^{-7} M), sodium nitroprusside (10^{-8} M) and papaverine (10^{-6} M) produced a moderate relaxation in the arterial strips denuded of the endothelium and contracted with $PGF_{2\alpha}$. In the strips treated with 3×10^{-6} M nitroglycerin, the relaxations induced by nicotine, NO and nitroprusside were almost abolished, whereas the response to papaverine was unaffected [3]. Nitroglycerin, NO and nitroprusside relax blood vessels by a mediation of cyclic GMP, but the papaverine-induced relaxation is not dependent on cyclic GMP. Excess production of cellular cyclic GMP by the high concentration of nitroglycerin seems to abolish the response mediated by cyclic GMP. We measured the content of cyclic GMP in endothelium-denuded arterial strips. Nicotine doubled the nucleotide content, compared to control [3]. The effect was abolished by hexamethonium and L-NA. These findings strongly suggest that the increased production of cyclic GMP by nicotine participates in the relaxation.

From endothelium-denuded cerebroarterial strips, nicotine markedly increased the release of NO_x [9]. Hexamethonium abolished the effect of nicotine. Similar findings were also obtained in the strips in which vasodilator nerves were electrically stimulated [9], although this effect was abolished by tetrodotoxin, instead of hexamethonium.

Nerve bundles and fibers containing NO synthase immunoreactivity were histologically demonstrated in dog cerebral arteries by the use of NO synthase antiserum raised against rat cerebellum, provided by Drs. Bredt and Snyder. There were abundant immunoreactive fibers and bundles forming many networks around the artery in whole mount preparations. In order to observe precise location of positive fibers, cryostat sections were cut slightly oblique to the coronal plane.

Plenty of nerve bundles and fibers were mainly situated in the adventitia, but fine fibers were also observed in the outer layer of the media [10].

The findings obtained from the functional study with nicotine and from the histochemical study strongly support the hypothesis that NO acts as neurotransmitter of vasodilator nerves innervating the dog cerebral arterial wall. We therefore call this nerve "nitroxidergic" [11]. All of the data presented here fit the known criteria for the neurotransmitter hypothesis, except for the fact that NO itself is not present in nerve endings, whereas neurotransmitters so far we know up to now are stored and released by nerve activation. Our hypothesis on this problem is that NO is synthesized from L-arginine by NO synthase which is activated by Ca^{2+} intracellularly introduced by nicotine or nerve action potentials. Therefore, we do not think that NO itself is necessarily present in nerve endings.

In conclusion, NO is synthesized in nerve endings innervating cerebral arteries from L-arginine by NO synthase which is activated by Ca^{2+}. The nerve is originated from the pterygopalatine ganglion [12]. NO liberated from the nerve as neurotransmitter activates soluble guanylate cyclase in smooth muscle and increases the production of cyclic GMP, resulting in the relaxation.

Substance P and vasopressin

These peptides produced moderate or marked relaxations of dog basilar and middle cerebral arterial strips with the intact endothelium, partially contracted with $PGF_{2\alpha}$. In contrast to nicotine, the response was endothelium-dependent.

Vasopressin-induced relaxations differed in cerebral arteries of various regions [13]. All of basilar and posterior cerebral arterial strips used responded to the peptide with moderate relaxations. On the other hand, only about half of middle cerebral arterial strips used responded with slight relaxations, but the peptide elicited contractions in the remaining strips. The relaxations induced were endothelium-dependent. According to Katusic et al. [14], the V_1-vasopressinergic receptor subtype is involved in the relaxation.

Relaxations induced by 10^{-8} M substance P and 10^{-8} M vasopressin were not influenced by indomethacin. Treatment with L-NA attenuated the relaxations in a dose-dependent manner. The vasopressin-induced relaxation was reversed to a contraction by high concentrations of the NO synthase inhibitor. D-NA did not inhibit the response [15]. The relaxation induced by NO was not inhibited by L-NA. The peptide-induced relaxations inhibited by L-NA were partially reversed by L-arginine. D-arginine did not restore the response. It appears that relaxations associated with substance P and vasopressin are mediated exclusively by NO derived from the endothelium in dog cerebral arteries, despite the fact that endothelium-dependent relaxations of some other blood vessels are reported to derive partially from mediators other than NO [16–18]. Treatment with cadmium significantly attenuated the relaxations caused by the peptides, as seen in the response to nicotine, whereas nicardipine was ineffective [8]. Together with the data reported elsewhere on cultured endothelial cells in which persistent increase in

44

intracellular Ca^{2+} by bradykinin is suppressed by polyvalent cation Ca^{2+} entry blockers [19,20], we speculate that the inhibition by cadmium of the response to EDRF-releasing peptides is associated with the interference with Ca^{2+} influxes through non-L-type Ca^{2+} channels.

In the endothelium-intact cerebral arterial strips, substance P increased the content of cyclic GMP [15]. Oxyhemoglobin and L-NA abolished the effect of the peptide and also depressed the content of cyclic GMP to the level attained by endothelium denudation that was about 10% of control.

NO is synthesized from L-arginine by constitutive NO synthase in the perivascular nerve and endothelium that is activated by Ca^{2+}. Even under resting conditions in vivo, the arterial smooth muscle tone would be regulated by NO released from nerves placed under tonic efferent influences from the central nervous system and by basal release of NO from endothelial cells. We wanted to determine and distinguish such a contribution of the nerve and endothelium to the regulation of dog basilar arterial tone in vivo.

Nerve- and endothelium-derived NO in basilar arteries in vivo

Again, we angiographically measured the basilar arterial diameter. Five ml of the cerebrospinal fluid was removed, and a test drug in the same volume was slowly injected into the subarachnoid space. Sixty min after intracisternal injection of L-NA, the basilar artery constricted. Intracisternal L-arginine reversed the induced vasoconstriction [1]. The time course of the effects is shown in Fig. 3. The Hartman's solution, a solvent of L-NA, in the same volume did not alter the basilar arterial diameter.

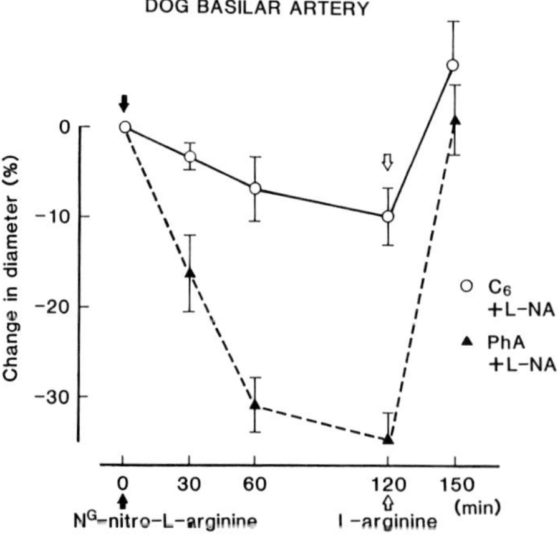

Fig. 3. Time course of effects of L-NA and L-arginine on basilar arterial diameter angiographically measured in anesthetized dogs treated with hexamethonium (C_6) or phentolamine (PhA) which produced the similar magnitude of hypotension. Partially reproduced from Toda et al., Am J Physiol 1993; 265: H103–H107, #1.

The ordinate indicates percent changes in the diameter relative to that before the application of L-NA. D-NA did not alter the arterial diameter. Treatment with hexamethonium markedly reduced the vasoconstrictor effect of L-NA. These findings suggest that the magnitude of attenuation caused by hexamethonium of the vasoconstrictor response to L-NA is associated with a suppression of nitroxidergic nerve activity. Tonic nitroxidergic innervation appears to play an important role in dilating basilar arteries. The L-NA-induced vasoconstriction still observed in hexamethonium-treated dogs is associated with NO synthesized in extraneuronal tissues, mainly in the endothelium. It is concluded that under the experimental conditions so far used, NO derived from vasodilator nerves contributes to the regulation of basilar arterial tone predominantly over NO derived from extraneuronal tissues.

References

1. Toda N, Ayajiki K, Okamura T. Am J Physiol 1993;265: H103–H107.
2. Suzuki Y, Satoh S, Oyama H, Takayasu M, Shibuya M. Stroke 1993;24:1049–1054.
3. Toda N, Okamura T. J Pharmacol Exp Ther 1991;258:1027–1032.
4. Toda N. J Pharmacol Exp Ther 1975;193:376–384.
5. Toda N. Am J Physiol 1982;243: H145–H153.
6. Toda N. J Cerebral Blood Flow & Metab 1988;8:46–53.
7. Toda N, Okamura T. Am J Physiol 1990;259: H1511–H1517.
8. Toda N, Okamura T. J Pharmacol Exp Ther 1992;261:234–239.
9. Toda N, Okamura T. Biochem Biophys Res Commun 1990;170:308–313.
10. Yoshida K, Okamura T, Kimura H, Bredt DS, Snyder SH, Toda N. Brain Res 1993;629:67–72.
11. Toda N, Okamura T. News Physiol Sci 1992;7:148–152.
12. Toda N, Ayajiki K, Yoshida K, Kimura H, Okamura T. Circ Res 1993;72:206–213.
13. Onoue H, Nakamura N, Toda N. Stroke 1988;19:1388–1394.
14. Katusic ZS, Shepherd JT, Vanhoutte PM. J Pharmacol Exp Ther 1986;236:166–170.
15. Toda N, Ayajiki K, Okamura T. J Vasc Res 1993;30:61–67.
16. Cowan CL, Cohen RA. Am J Physiol 1991;261: H830–H835.
17. Matsumoto T, Kinoshita M, Toda N. J Cardiovasc Pharmacol 1993;21:228–234.
18. Enokibori M, Okamura T, Toda N. Br J Pharmacol 1994;111:77–82.
19. Colden-Stanfield M, Schilling WP, Ritchie AK, Eskin SG, Nazarro LT, Kunze DL. Circ Res 1987; 61:632–640.
20. Buchan KW, Martin W. Br J Pharmacol 1991;102:35–40.

Endothelium-Derived Factors and Vascular Functions. T. Masaki, ed.

Endothelial NO and vascular regulation

Michel Félétou[2], Emmanuel Canet[2] and Paul M. Vanhoutte[1]
[1]Direction Recherche et Développement, IRIS, 6 place des pléiades, 92415 Courbevoie;
*[2]Département de Pneumologie, Institut de Recherches Servier, 11 rue des Moulineaux, 92150
Suresnes, France*

Introduction

The endothelium is at the interface between the blood and the underlying vascular
cells and occupies a strategic position in the control of local hemodynamics. It
modulates the tone of vascular smooth muscle by acting as a diffusion barrier, by
degrading or activating circulating substances, and by producing vasoactive factors.
It also controls vascular permeability and interacts with the blood cells to regulate
blood fluidity and hemostasis.

In 1980, Furchgott and Zawadzki [1] demonstrated that the relaxation of isolated
arteries to acetylcholine requires the presence of endothelial cells. This response is
mediated by a labile humoral substance termed endothelium-derived relaxing factor
(EDRF). Since then, many investigators have repeated this seminal observation in
a variety of blood vessels with a number of vasoactive agents (for review see [2–6]).
EDRF causes relaxation of vascular smooth muscle by activation of soluble
guanylate cyclase and increasing intracellular level in cyclic guanosine monophosphate (cyclic GMP), a mechanism of action shared with exogeneous nitrovasodilators [7–9]. Based on their very similar pharmacological profile, Furchgott [10] and
Ignarro et al. [11] proposed that EDRF is the nitric oxide radical (NO) or a closely
related compound. Moncada and coworkers then demonstrated that NO, which is
released from endothelial cells, is synthesized from molecular oxygen and one of
the terminal guanido nitrogen atoms of L-arginine [12-14]. Nitric oxide synthase, the
enzyme responsible for NO production, is an FAD and FMN containing enzyme
which requires NADPH and activation of the calcium-calmodulin pathway for
maximal activity [15–17] (Fig. 1). This enzyme has a close homology to another
mammalian electron-transferase enzyme, cytochrome P450 reductase. It is expressed
constitutively not only in endothelial cells but also in platelets and in the central and
peripheral nervous system. A calcium-independent form of the nitric oxide synthase
can be induced by different stimuli (e.g.: lipopolysaccharide, cytokines...) in
numerous cells (for review see [16–20]). In humans, three different genes code for
the endothelial (chromosome 7), neuronal (chromosome 12) and inducible
(chromosome 17) nitric oxide synthase. The endothelial enzyme has been cloned
and characterized [21,22].

NO is not the only vasoactive factor synthesized and released by the endothelial
cells. Thus, adenosine [23], prostaglandins [24], an unidentified hyperpolarizing

52

Cardiovascular pathology and endothelial NO

Hypertension

Acute increases in blood pressure cause endothelial damage, and a loss of endothelium-dependent relaxation in the cerebral vascular bed [74,75]. In animal models of chronic hypertension, endothelial cells exhibit morphological changes and the response of isolated blood vessels to endothelium-dependent dilators is reduced [76,77]. In genetically spontaneously hypertensive rats, this reduction is not related to a diminution of NO production but to that of endothelium-derived endoperoxide(s), which directly interacts with NO [78,79]. In most of the other models of hypertension (e.g: coarctation, mineralocorticoid hypertension, renal hypertension, salt sensitive Dahl rat), the impairment of endothelium-dependent relaxation is restricted to some vascular beds and varies according to the hypertension model studied [80]. A deficit in NO production, a decrease in the sensitivity of the smooth muscle to NO, the production of constricting factors which counterbalances the vasodilator action of NO or directly inactivates it, all can contribute to the vascular dysfunction. The endothelial impairment seems to be a consequence rather than a primary cause of hypertension since antihypertensive treatments normalize endothelium-dependent relaxations [80-82]. However in Dahl rats, high dietary potassium reduces the incidence of stroke and augments endothelium-dependent relaxations; these two effects are independent of blood pressure [82,83].

In humans, nitric oxide-dependent dilator responses seem to be abnormal in hypertension. Thus, a NO synthase inhibitor (LNMMA) is less effective in decreasing forearm blood flow in hypertensive patients than in control subjects [84,85]. Furthermore, the vasodilator response to acetylcholine, which is specifically reduced in hypertensive patients compared to control subjects, is not significantly affected by LNMMA in the former [85]. This phenomenon is independent of substrate availability as L-arginine infusion does not modify the response to acetylcholine in hypertensive subjects [86]. There is also some evidence for a reduction in platelet nitric oxide synthase in essential human hypertension [87]. This is unfavorable since the protective function of the endothelium against platelet aggregation and adhesion, and platelet-induced contraction is impaired [88,89]. These alterations in endothelium-platelet interaction may help to explain vascular complications such as ischemic stroke, myocardial infarction and peripheral vascular diseases [76].

Even mild impairment of NO synthesis in the systemic circulation produces a decrease in sodium excretion which in turn favors an increase in blood pressure during high sodium intake. Deficiency in the synthesis of nitric oxide could be involved in the development of salt-sensitive hypertension [90]. The immunosuppressant agent, cyclosporin A, induces nephrotoxicity and hypertension. This deleterious effect could be due to an impairment of endothelial function [91]. In humans, an endogeneous inhibitor of the NO synthase, asymmetrical dimethyl-arginine (ADMA), is produced and excreted in the urine. In patients with end-stage chronic renal failure, urine output is markedly decreased and there is sufficient accumulation of ADMA in the plasma to block endogeneous NO production. This could contribute to the hypertension associated with this pathology [92].

The effects of a number of antihypertensive agents is affected by the presence of the endothelial cells. The non-selective β-adrenergic blocker, carteolol, potentiates endothelium-dependent relaxations to $\alpha2$ adrenergic activation [93]. Other non-selective β-adrenergic blockers may induce endothelium-dependent relaxation [94]. Nebivolol, a $\beta1$-adrenergic antagonist, induces NO-dependent relaxation unrelated to its β-adrenergic blocking properties [95]. This implies that part of the unexplained vasodilator effects of β-adrenergic blockers may be due to facilitation of endothelium-dependent relaxation of arterial smooth muscle. S11568, a dihydropyridine calcium channel blocker, and its optical isomers evoke NO-dependent relaxations; this effect is distinct from the calcium antagonistic action which is stereoselective [96]. Amlodipine, a chemically related calcium antagonist, induces inhibition of platelet aggregation through an NO mediated process [97]. In contrast d-cis diltiazem and its analogue TA3090, two benzothiazepine calcium antagonists, inhibit the synthesis or the release of endothelium-derived relaxing factor [98,99].

Angiotensin converting enzyme (ACE) inhibitors protect bradykinin from degradation which results in greater production of nitric oxide and hyperpolarizing factor, via stimulation of endothelial B2 kinin receptors [68,69,100]. The dilatation evoked by ACE inhibitors in hypertensive patients is related to protection of bradykinin [101]. Bradykinin could also be involved, through an endothelial NO pathway, in ischemic preconditioning of the heart, which prevents occlusion-reperfusion induced arrhythmia or sudden death, indeed converting enzyme inhibitors potentiate the beneficial effect of preconditioning [101].

Hypercholesterolemia and atherosclerosis

A reduced endothelium-dependent relaxation occurs very early in atherosclerosis [103]. For example, in vivo denudation of the endothelium of porcine coronary arteries is followed by a rapid regeneration of the intimal layer. After endothelial regeneration, the endothelium-dependent responses to various agonists (bradykinin, ADP, thrombin...) are preserved; however, the endothelium-dependent relaxation to serotonin and $\alpha2$-adrenoceptor agonists is attenuated, and the endothelial response to aggregating platelets is impaired [104,105]. This reflects the loss in the regenerated endothelium of a pertussis toxin sensitive pathway linking membrane receptors to NO synthase (Fig. 4) [105].

Oxidized low density lipoprotein (LDL), but not native LDL (the major carrier of cholesterol in the blood), inhibits endothelium-dependent relaxation in vitro [106]. Arteries taken from pigs and rabbits fed a hypercholesterolemic diet exhibit reduced endothelium-dependent relaxations [107–109]. In rabbits, in a later stage of atherosclerosis, the reduction of endothelium-dependent relaxation is related to the level of cholesterol, the duration of the diet and the degree of fatty streak formation [110,111].

In the pig, denudation of the coronary artery endothelium, followed by an hypercholesterolemic diet results in the rapid development of atherosclerosis. In

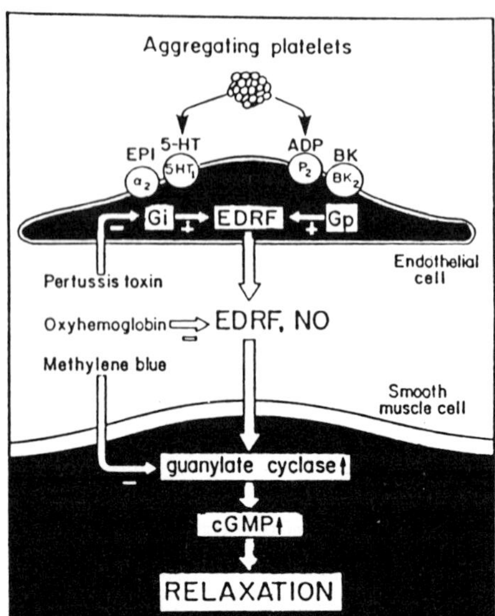

Fig. 4. Release of endothelium-derived relaxing factor (EDRF) and G protein transduction. The release of NO by endothelial cells involves at least two types of G proteins, one of which is sensitive to pertussis toxin. This pathway is affected by endothelial regeneration and atherosclerosis. ADP = adenosine diphosphate; BK = bradykinin; EP1 = epinephrine; 5-HT = 5-hydroxytryptamine; α2 = adrenoceptor, B2 = bradykinin receptor; 5-HT1 = serotonin receptor; P2 = purinergic receptor; cGMP = cyclic guanosine monophosphate; Gi and Gp = Gi and Gp proteins [140].

these blood vessels, the endothelium-dependent relaxation to serotonin is blunted but the responses to bradykinin, ADP, thrombin are impaired also [112]. The loss of the endothelial pertussis toxin-sensitive pathway in response to serotonin and α2-adrenergic agonists is particularly obvious [113]. The reduced endothelium-dependent relaxation in atherosclerotic arteries involves a reduced release of NO, but other factors are present as the disease progresses: superoxide anion generation which scavenges NO and a reduced sensitivity of the smooth muscle to the endothelial factor [105,110]. In human atherosclerosis, the impairment of endothelial function has been demonstrated both in vitro and in vivo (e.g. [103,114-116]).

Epidemiologic studies have shown that a diet rich in marine oil prevents the development of coronary artery disease and lowers the incidence of atherosclerosis [117]. In pigs fed with a diet containing unsaturated fatty acid (eicosapentaenoic acid and docasahexaenoic acid), the endothelium-dependent relaxations to aggregating platelets are potentiated. This effect can be demonstrated in healthy, hypercholesterolemic and atherosclerotic animals [118,119].

Reperfusion injury

Coronary vasospasm can occur after reperfusion of the previously ischemic myocardium [120]. In coronary arteries, studied ex vivo after occlusion-reperfusion, an acute impairment of endothelium-dependent relaxations to various neuro-humoral substances and to aggregating platelets is observed [121-124]. This early endothelial dysfunction can be prevented by superoxide dismutase, indicating the involvement of superoxide free radicals production [125]. The reperfusion injury is also characterized by the extensive adhesion of neutrophils to the endothelial cells. A decrease in basal release of endothelial NO, after myocardial ischemia-reperfusion, may, per se, promote the neutrophils adherence to the coronary artery endothelium [126]. In dogs, even twelve weeks following occlusion-reperfusion injury, a selective impairment of the endothelium-dependent relaxation to aggregating platelets is still present [127]. In vivo this phenomenon could favor vasospasm and coronary artery thrombosis.

Diabetes

In animal models of diabetes mellitus, alterations in vascular responsiveness, and especially an impairment of endothelium-dependent responses is observed [128-130]. The reduced endothelium-dependent relaxation is not related to a decrease in NO production but to the generation of endothelial endoperoxide/thromboxane which counterbalances the effect of NO, or inactivates it [128,131].

In human diabetes, dysfunctional endothelial responses can also be demonstrated. Thus the infusion of LNMMA is less effective in diminishing the forearm blood flow in patients with diabetes than in control subjects. This abnormality is at least partly related to a decreased sensitivity of the smooth muscle to basally released NO [132].

In the aging rat developing hyperinsulinemia and insulin-resistance (type II diabetes), a decrease in the endothelium-dependent relaxation to ADP occurs combined with hyperreactivity of the vascular smooth muscle to serotonin [133]. Chronic treatment with D-fenfluramine, a drug which reduces insulinemia, normalizes these two abnormalities. In the genetically obese hyperinsulinemic Zucker rat, an hyperreactivity to serotonin is present also [134]. These alterations in the response to platelet products may contribute to vasospasm and thrombosis in patients with syndrome X (insulin-resistance, glucose tolerance, hypertension [135]).

Subarachnoid hemorrhage

Various clinical and experimental studies of cerebral arteries in chronic vasospasm have demonstrated structural damage to each layer of the arterial wall including the endothelium. In a canine model of subarachnoid hemorrhage, endothelium-dependent relaxations are suppressed while endothelium-dependent contractions are preserved [136]. This alteration of endothelial function could be of importance in the pathogenesis of cerebral vasospasm. However, impairment of endothelium-dependent relaxations is not related to a decrease in NO production, but to a reduced activation of the soluble guanylate cyclase of the vascular smooth muscle, or to a decreased transfer of the radical to the cells possibly because of trapping by hemoglobin [136,137].

Circulatory shock

Among the more common shock-induced stimuli are: trauma, severe hemorrhage, endotoxemia, splanchnic ischemic shock and cardiogenic shock. All these conditions lead to severe splanchnic hypoperfusion and impairment of endothelium-dependent relaxation. Super oxide radical, that scavenges NO [138], plays a major role in this phenomenon [139]. In septic shock, the expression of an inducible nitric oxide synthase is probably involved in the second phase of the fall in blood pressure [6].

References

1. Furchgott RF, Zawadzki JV. Nature 1980;288:373–376.
2. Furchgott RF, Vanhoutte PM. Faseb J 1989;3:2007–2018.
3. Feletou M, Vanhoutte PM. J Nephrol 1990;1:17–27.
4. Luscher TF, Vanhoutte PM. The endothelium: modulator of cardiovascular function. Boca Raton Fla: CRC Press, Inc 1990;1–228.
5. Bassenge E, Heusch G. Rev Physiol Biochem Pharmacol 1990;116:77165.
6. Moncada S, Palmer RMJ, Higgs EA. Pharmacol Rev 1991;43(2):109–142.
7. Rapoport RM, Murad F. Circ Res 1983;52:52–57.
8. Ignarro LJ, Kadowitz PJ. Ann Rev Phannacol Toxicol 1985;2:171–191.
9. Forstermann U, Mulsch A, Bohme E, Busse. Circ Res 1986;58:531–538.
10. Furchgott RF. In: Vanhoutte PM (Ed.) Mechanism of Vasodilatation, vol 4, New York: Raven Press, 1988;401–414.
11. Ignarro JL, Byrns RE, Wood KS. In: Vanhoutte PM (Ed.) Mechanism of Vasodilatation, vol 4, New York: Raven Press, 1988;427–435.
12. Palmer RMJ, Ferridge AG, Moncada S. Nature 1987;327:524–526.
13. Palmer RMJ, Ashton DS, Moncada S. Nature 1988;333;664–666.
14. Palmer RMJ, Rees DD, Ashton DS, Moncada S. Biochem Biophys Res Comm 1988;153:1521–1526.
15. Bredt DS, Snyder SH. Proc Nat Acad Sci, USA 1990;87:682–685.
16. Moncada S. Acta Physiol Scand 1992;145:201–227.
17. Schini V, Vanhoutte PM. J Pharmacol Exp Ther 1992;261(2):553–559.
18. Nathan C. FASEB J 1992;6:3051–3064.
19. Toda N, Okamura T. News Physiol Sci 1992;7:148–152.
20. Snyder S. Science 1992;257:494–496.
21. Marsden PA, Schappert KT, Chen HS et coll. Febs letters 1992;307(3):287–293.
22. Janssens SP, Shimouchi A, Quertemou T, Bloch PR, Bloch KD. J Biol Chem 1992;267(21):14519–14522.
23. Engler RL, Gruber IIE. In: Fozzard HA et al. (Ed.) The heart and cardiovascular system, 2nd edn. New York: Raven, 1991;1745–1764.
24. Moncada S, Vane JR. Pharmacol Rev 1979;30:293–331.
25. Feletou M, Vanhoutte PM. Br J Pharmacol 1988;93:515–522.
26. Vanhoutte PM. Circulation 1993;87(suppl V):9–17.
27. Yanagisawa M, Kurihara H, Kimura S. Nature 1988;332:411–415.
28. Luscher PM, Boulanger CM, Dohi Y, Yang Z. Hypertension 1992;19:117–130.
29. Pohl U, Busse R, Kuon E, Bassenge E. J Appi Cardiol 1986;1:215–235.
30. Rubanyi GM, Romero JC, Vanhoutte PM. Am J Physiol 1986;250: H1145–H1149.
31. Inoue T, Tomoike H, Hisano K, Nakamura M. J Am Coll Cardiol 1988;11:187–191.
32. Miller VM, Vanhoutte PM. Am J Physiol 1988;255: H446–H451.
33. Holtz J, Forztermann U, Giesler M, Bassenge E. J Cardiovasc Pharmacol 1984;6:1161–1169.

34. Kaiser L, Sparks HV. Circ Shock 1986;18:109–114.
35. Bevan JA, Joyce EH, Wellman GC. Circ Res 1988;6:980–985.
36. Harder D. Circ Res 1987;60:102–107.
37. Harder DR, Sanchez-Ferrer C, Kauser K, Stekiel WJ, Rubanyi GM. Circ Res 1989;65:193–198.
38. Katusic ZS, Shepherd JT, Vanhoutte PM. Am J Physiol 1987;252: H671–H673.
39. Rees DD, Palmer RMJ, Hodson HF, Moncada S. Br J Pharmacol 1989;96:418–424.
40. Rees DD, Palmer RMS, Moncada S. Proc Natl Acad Sci USA 1989;86:3375–3378.
41. Haynes WG, Noon JP, Walker BR, Werd DJ. Endothelium 1993;1(suppl S):14.
42. Stamler JS, Loh E, Roddy M, Hoffman KE, Creager MA. Endothelium 1993;1(suppl S):14.
43. Cocks TM, Angus JA. Nature 1983;305:627–630.
44. Miller VM, Vanhoutte PM. Eur J Pharmacol 1985;118:123–129.
45. Feletou M, Grau F, Teisseire B. Fondam Clin Pharmacol 1993;7:209–217.
46. Katusic ZS, Shepherd JT, Vanhoutte PM. Circ Res 1984;55:575–579.
47. Katusic Z, Moncada S, Vanhoutte PM. In: Moncada S, Higgs A (eds) Nitric oxide from L-arginine: a bioregulatory system. Amsterdam: Elsevier, 1990;69–72.
48. Vidal MJ, Romero JC, Vanhoutte PM. Eur J Pharmacol 1988;149:401–402.
49. Lahera V, Salom MG, Miranda F, Moncada S, Romero JC. Am J Physiol 1991;261: F1033–F1037.
50. Salazar FJ, Pinilla JM, Lopez F, Romero JC, Quesada T. Hypertension 1992;20:113–117.
51. Fukuda Y, Hirata Y, Yoshimi H et coll. Biochem Biophys Res Comm 1988;155:167–172.
52. Sanchez-Ferrer CF, Burnett JC, Lorenz RR, Vanhoutte PM. Am J Physiol 1990;259(28): H982–H986.
53. Boulanger C, Luscher TF. J Clin Invest 1990;85:587–590.
54. Azuma H, Ishikawa M, Sckizakis S. Br J Pharrnacol 1986;88:411–415
55. Radomski MW, Palmer RMJ, Moncada S. Biochem Biophys Res Comm 1987;148(3):1482–1489.
56. Cohen RA, Shepherd JT, Vanhoutte PM. Science 1983;221:273–274.
57. Vanhoutte PM, Houston DS. Circulation 1985;72(4):728–734.
58. Radomski MW, Palmer RMJ, Moncada S. Br J Pharmacol 1987;92:639–646.
59. Kamata K, Mori T, Shigenobu K, Kasuya Y. Br J Pharmacol 1989;9:1360–1364.
60. Erdos EG. In: Erdos EG (ed.) Handbook of experimental pharmacology, vol 25. New York: Springer Berlin Heidelberg, 1979;427–487.
61. Dzau VJ. Am J Med 1984;77:31–36.
62. Vanhoutte PM. Hypertension 1989;13:658–667.
63. Feletou M, Germain, Teisseire B. J Vasc Med Biol 1989;1(6):319–329.
64. Furchgott RF. Acta Physiol Scand 1990;139:257–270.
65. Toda N. Br J Pharmacol 1984;81:301–307.
66. Feletou M, Germain, Teisseire B. J Vasc Med Biol 1990;2(1):1825.
67. Feletou M, Gemain M, Teisseire B. Am J Physiol 1992;262: H839–H845.
68. Feletou M, Teisseire B. Eur J Pharmacol 1990;190:159–166.
69. Mombouli JV, Illiano S, Nagao T, Scott-Burden T, Vanhoutte PM. Circ Res 1992;71:137–144.
70. Mombouli JV, Vanhoutte PM. J Cardiovasc Pharmacol 1992;20(suppl 9): S74–S83.
71. Garg UC, Hassid A. J Clin Invest 1989;83:1774–1777.
72. Scott-Burden T, Vanhoutte PM. Circulation 1993;87(suppl V):51–55.
73. Mayhan WG. Inflammation 1992;16(4):295–304.
74. Lamping KG, Dole WP. Circ Res 1987;61:904–913.
75. Kontos HA. Circ Res 1958;57:508–516.
76. Luscher TF. In: Karger Barel S (ed.), Publisher AG, 1988;1–133.
77. Vanhoutte PM, Luscher TF. In: Tarazi RC (ed.) Handbook of Hypertension. Physiology and Pathophysiology of Hypertension – Regulatory mechanisms, vol 8. Amsterdam: Elsevier, 1987;96–123.
78. Luscher TF, Vanhoutte PM. Hypertension 1986;8:344–348.
79. Auch-Schwelk W, Katusic ZS, Vanhoutte PM. Hypertension 1992;19:442–445.
80. Luscher TF. Am J Hypertension 1990;3:317–330.
81. Van de Voorde J, Vanheel B, Leusen I. Pflügers Arch 1988;411:500–504.

82. Raij L, Luscher TF, Vanhoutte PM. Hypertension 1988;12:562–567.
83. Tobian L, Lange JM, Ulm KM, Wold LJ, Iwai J. J Hypertens 1984;2(suppl 8):363–366.
84. Calver A, Collier J, Moncada S, Vallance P. J Hypertens 1992;10(9):1025–1031.
85. Panza JA, Casino PR, Kilcoyne CM, Quyyumi AA. Circulation 1993;87(5):1468–1474.
86. Panza JA, Casino PR, Badar DM, Quyyumi AA. Circulation 1993;87(5):1475–1481.
87. Cadwgan TM, Benjamin N. J Hypertens 1993;11(4):417–420.
88. Hazama F, Ozaki T, Amano S. Stroke 1979;10:245–252.
89. Luscher TF, Vanhoutte PM. Hypertension 1986;8(suppl II):55–60.
90. Romero JC, Lahera V, Salom MG, Biondi ML. J Am Soc Nephrol 1992;2(9):1371–1387.
91. Gerkens JF. J Pharmacol Exp Ther 1989;250:1105–1112.
92. Vallance P, Leone A, Calver A, Collier J, Moncada S. Lancet 1992;339(8793):572–575.
93. Janczewski P, Boulanger C, Iqbal A, Vanhoutte PM. J Pharmacol Exp Ther 1988;247:590–595.
94. Mostaghim R, Maddox YT, Ramwell PW. J Pharmacol Exp Ther 1986;239:797–801.
95. Gao Y, Nagao T, Bond RA, Janssens WJ, Vanhoutte PM. J Cardiovasc Pharmacol 1991;17: 964–969.
96. Vilaine JP, Biondi ML, Villeneuve N, Feletou M, Peglion JL, Vanhoutte PM. Eur J Pharmacol 1991;197:41–48.
97. Berkels R, Klaus W, Rosen R. Endothelium 1993;1(suppl S):69.
98. Rubanyi GM, Hoeffner U, Schwartz A, Vanhoutte PM. J Pharmacol Exp Ther 1988;246(1):60–64.
99. Rubanyi GM, Ibqal A, Schwartz A, Vanhoutte PM. J Pharmacol Exp Ther 1991;259(2):639–642.
100. Vanhoutte PM, Auch-Schwelk W, Biondi MLS, Lorenz RR, Schini VB, Vidal MJ. Br J Clin Pharmacol 1989;28:95–104.
101. Gavras H, Gavras I. Am J Med Sci 1988;295:305–307.
102. Parratt J. Cardiovasc Res 1993;27(5):693–702.
103. Mc Lenachan JM, Vita J, Fish RD, Treasure CB, Cox DA, Ganz P, Selwyn AP. Circulation 1990;82:1169–1173.
104. Shimokawa H, Aarhus LL, Vanhoutte PM. Circ Res 1987;61:256–270.
105. Shimokawa H, Flavahan NA, Vanhoutte PM. Circ Res 1989;65:740–753.
106. Simon BC, Cunningham LD, Cohen RA. J Clin Invest 1990;86:75–79.
107. Cohen RA, Zitnay KM, Haudenschild CC, Cunningham LD. Circ Res 1988;63:903–910.
108. Shimokawa H, Vanhoutte PM. J Am Coll Cardiol 1989;13:1402–1408.
109. Osborne JA, Segman MJ, Sedar AW, Mooers SU, Lefer AM. Am J Physiol 1989;256: C591–C597.
110. Verbeuren TJ, Jordaens FH, Zonnekeyn LL, Vanhove CE, Coene MC, Herman AG. Circ Res 1986;58:552–564.
111. Verbeuren TJ, Herman AG. In: Vanhoutte PM (ed.) Relaxing and contracting factors: biological and clinical research. Clifton NJ: Humana Press, 1988;451–472.
112. Shimokawa H, Vanhoutte PM. Circ Res 1989;64:900–914.
113. Shimokawa H, Flavahan NA, Vanhoutte PM. Circulation 1991;83:652–660.
114. Forstermann U, Mugge A, Alheid U, Haverich A, Frolich JC. Circ Res 1988;62:185–190.
115. Cox DA, Vita JA, Treasure CB, Fish RD, Alexander WR, Ganz P, Selwyn AP. Circulation 1989;80:458–465.
116. Hasue H, Matsuyama K, Okumura K, Morikami Y, Ogawa H. Circulation 1990;81:482–490.
117. Bang HO, Dyerberg J. Adv Nutr Res 1980;3:1–22.
118. Shimokawa H, Vanhoutte PM. Am J Physiol 1988;256: H968–H973.
119. Shimokawa H, Vanhoutte PM. Circulation 1988;78:1421–1430.
120. Dorros G, Cowley MJ, Simpson et coll. Circulation 1983;67:723–730.
121. Ku DD. Science 1982;218:576–578.
122. Vanbenthuysen KM, Mc Murtry IF, Horwitz LD. J Clin Invest 1987;79:265–274.
123. Mehta JL, Nichols WW, Lawson DL, Saldeen TGP. Circ Res 1989;64:43–54.
124. Pearson PJ, Schaff HV, Vanhoutte PM. Circ Res 1990;67:385–393.
125. Tsao PS, Lefer AM. Am J Physiol 1990;259: H1660–H1666.
126. Ma XL, Weyrich AS, Lefer DJ, Lefer AM. Circ Res 1993;72(2):403–412.
127. Pearson PJ, Schaff HV, Vanhoutte PM. Circulation 1990;81:1921–1927.

128. Tesfamarian B, Jabukowski JA, Cohen RA. Am J Physiol 1989;257: H1327–H1333.
129. Kamata K, Miyata N, Abiau T, Kasuya Y. Life Sciences 1992;50:1379–1387.
130. Feletou M, Moreau N, Duhault J. Life Sciences, In press 1993.
131. Feletou M, Rasetti C, Duhault J. J Vasc Res 1992;29:113.
132. Calver A, Collier J, Vallance P. J Clin Invest 1992;90(6):2548–2554.
133. Feletou M, Moreau N, Boulanger M, Duhault J. J Cardiovasc Pharmacol 1993;21:120–127.
134. Zemel MB, Reddy S, Shehin SE, Lockette W, Sowers JR. J Vasc Med Biol 1990;2(2):81–85.
135. Reaven GM. Diabetes 1988;37:1595–1607.
136. Kim P, Sundt TM, Vanhoutte PM. J Neurosurg 1988;69:239–246.
137. Kim P, Lorenz RR, Sundt TM, Vanhoutte PM. Cir Res 1992;70:248–256.
138. Rubanyi GM, Vanhoutte PM. Am J Physiol 1986;250: H222–H227.
139. Lefer AM, Lefer DF. Ann Rev Pharmacol Toxicol 1990;33:71–90.
140. Vanhoutte PM. Eur Heart J 1991;12(suppl E):25–32.

Endothelium-Derived Factors and Vascular Functions. T. Masaki, ed.

Mechanisms of endothelin-induced contraction and relaxation of vascular smooth muscle

H. Kanaide

Division of Molecular Cardiology, Research Institute of Angiocardiology, Faculty of Medicine, Kyushu University, Fukuoka 812, Japan

Summary

Prior to prolonged vasoconstriction, endothelin-1 induces a transient vasodilatation when injected in vivo. Endothelin-1 induces a transient elevation of (Ca)i in endothelial cells and the release of EDRF. In addition to ET_B receptors, there are ET_A receptors in endothelial cells, which may play an important role when endothelin-1 stimulates endothelial cells to release EDRF, in situ. Thus, endothelin-1 may play an important role in controlling arterial tone not only by acting directly on smooth muscle cells to increase tension in a paracrine manner, but also by acting on endothelial cells to release EDRF in an autocrine manner.

Introduction

Endothelin (ET) is a potent vasoconstrictor peptide and has at least three isoforms, ET-1, ET-2 and ET-3 [1,2]. The order of the potency as a vasoconstrictor is ET-2 ≥ ET-1 >> ET-3 [2]. As ET-1 is the only one of the three made by endothelial cells, it may directly contract the underlying vascular smooth muscle cells in vivo. In addition, when injected in vivo, prior to a prolonged vasoconstriction, ET-1 induced a transient, dose-dependent vasodilator response which was thought to be induced by endothelium-derived relaxing factor (EDRF) or by prostacyclin released by ET-1 from endothelial cells [3,4]. However, the precise mechanism of the vasodilating effects of ET-1 is a matter of controversy. In the present report, the mechanisms of the initial vasodilating effect and the vasoconstricting effect induced by ET-1 were investigated in vitro. Using front-surface fluorometry and fura-2 loaded porcine aortic valvular strips and coronary arterial strips with intact endothelial cells, the effect of ET on cytosolic Ca concentration, (Ca)i, in smooth muscle cells and endothelial cells were determined. Both the contraction of smooth muscle cells and the release of EDRF or prostacyclin from endothelial cells are regulated by the increase in (Ca)i.

62

Materials and Methods

Preparation of the aortic valvular strips and coronary arterial strips

Hearts and aortic roots from adult pigs of either sex were obtained from a local slaughterhouse immediately after the animal had been killed. The aortic valve leaflets were cut with special care not to touch their surface. The leaflets were cut in parallel with midline to make approximately 2 mm (width) × 5 mm (length) × 0.18 mm (thickness) valvular strips [5].

Left circumflex coronary arteries were isolated and segments 2–3 cm from the origin were excised. After removal of the adventitia, excised segments were cut into approximately 1 mm (width) × 5 mm (length) × 0.1 mm (thickness) circular strips. When indicated, the luminal surface of vascular strips were rubbed off with a cotton swab to remove the endothelium [6].

Fura-2 loading to valvular and vascular strips

The valvular strips were incubated for 1.5 hr (37°C) in oxygenated (5% CO_2, 95% O_2) Dulbecco's modified Eagle's medium containing 5% fetal bovine serum, 1 mM probenecid and 50 µM fura-2/AM. After loading the fura-2, the strips were washed 5 times by normal physiological salt solution (normal PSS: NaCl 123, KCl 4.7, $NaHCO_3$ 15.5, KH_2PO4 1.2, $MgCl_2$ 1.2, $CaCl_2$ 1.25 and D-glucose 11.5, in mM) and left for 1 hr at room temperature for purposes of equilibration. The measurements were carried out at 25°C to prevent the leakage of fluorescence dye [5]. Microscopic observations revealed that no cellular component other than endothelial cells had fura-2 fluorescence in aortic valvular strips, indicating that (Ca)i signals were exclusively from endothelial cells in situ.

The coronary arterial strips with or without endothelium were loaded with fura-2 by incubating in oxygenated Dulbecco's modified Eagle's medium containing 25 µM fura-2/AM and 2.5% fetal bovine serum for 3–4 hr (37°C). The strips were then rinsed with normal PSS for purposes of equilibration. The measurements were carried out at 37°C. Microscopic observations revealed that the fura-2 fluorescence of endothelial cells, if any, was negligible, and that changes in fluorescence indicated changes in (Ca)i in the smooth muscle cells.

Both the valvular strips and the arterial strips were mounted vertically in a quartz organ bath. Isometric tension development of the arterial strips was measured using a strain gauge transducer (TB-612T, Nihon Koden, Tokyo, Japan), simultaneously with (Ca)i.

Front-surface fluorometry

Changes in (Ca)i were monitored using a front-surface fluorometer (CAM-OF-1; designed in collaboration with Japan Spectroscopic Co., Tokyo, Japan), as described elsewhere [7]. In brief, the aortic surface of the valvular strip or the endothelial side of the coronary arterial strip was illuminated by guiding the alternating (400 Hz),

340 nm and 380 nm, excitation light from a xenon light source through quartz optic fibers arranged in a concentric inner circle (diameter: 3 mm). Surface fluorescence of the strip was collected by glass optic fibers arranged in an outer circle (diameter: 7 mm) and introduced through a 500 nm band-pass filter into a photon-counting photomultiplier. Fluorescence intensities (500 nm) at 340 nm (F340) and 380 nm (F380) excitation and its ratio (R = F340/F380) were continuously monitored. In case necessary, (Ca)i was calculated using the method described by Grynkiewicz et al. [8] with minor modifications. The apparent dissociation constants (K_d) of the fura-2:Ca complex at 25°C and 37°C were 162 nM and 224 nM, respectively, determined spectroscopically.

Normalization of fluorescence signals and developed force

Endothelial cells in the aortic valve: The responses of fluorescence ratio [F340/F380; (Ca)i] were normalized for those obtained with 10^{-5} M ATP, which was applied at least 15 min prior to each experimental measurement for 1 min. The resting level and the peak level of the fluorescence ratio induced by ATP were designated 0% and 100%, respectively. Absolute values of (Ca)i levels observed during 10^{-5} M ATP application were determined separately, and (Ca)i levels at 0% and 100% were 65.3 nM and 186 nM, respectively.

Smooth muscle cells in the coronary artery: Both the developed force and the fluorescence ratio [F340/F380; (Ca)i] were expressed as a percentage, designating the steady-state levels at rest (5.9 mM K^+) and high external K^+-depolarization (118 mM K^+ PSS) to be 0% and 100%, respectively. Absolute values of (Ca)i levels at 0% and 100% were 108 nM and 715 nM, respectively.

Reverse transcription polymerase chain reaction (RT-PCR) of porcine ET_A receptor in the endothelial cells on the aortic side of the porcine aortic valves

The nucleotide sequence of porcine lung ET_A receptor was determined using PCR technique. The similarities between porcine and bovine cDNA [9] sequences were 92% in the coding regions. The deduced amino acid sequence of porcine ET_A receptor was 97% identical to that of the bovine ET_A receptor. The specific primers were chosen according to this sequence. Total RNA prepared from the endothelial cells on the aortic side of the porcine aortic valves was first reverse-transcribed by 3′-primer, and then, applied to PCR amplification. A part of the PCR reaction mixture was applied to agarose gel electrophoresis (submitted for publ.).

Solutions and chemicals

The Ca-free solution (Ca-free PSS) contained 2 mM EGTA instead of 1.25 mM $CaCl_2$, and when indicated, contained 2 mM ethyleneglycol-bis(ß-aminoethylether)-*N,N,N′,N′*-tetraacetic acid (EGTA). Synthetic ET-1 and ET-3 were obtained from Peptide Institute Co. Ltd. (Osaka, Japan), and pertussis toxin (IAP) and fura-2/AM were purchased from Seikagaku Kogyo Co. (Tokyo, Japan) and Dojindo (Kumamoto, Japan), respectively. BQ-123 was donated from Banyu Pharmaceutical Co., Ltd. (Tokyo, Japan). All other reagents were of the highest grade available commercially.

Statistical analysis

Values were expressed as mean ± standard error. Student's *t*-test was used to determine statistical significance and *P* values less than 0.05 were considered to be significant.

Results

Changes in (Ca)i induced by ET-1 and ET-3 in endothelial cells in situ

Figure 1 shows representative recordings of changes in F340, F380 and the fluorescence ratio in endothelial cells in aortic valve induced by ET-1 and ET-3 in the presence and the absence of extracellular Ca [5]. When 10^{-6} M ET-1 was applied in the presence of extracellular Ca, (Ca)i rapidly rose to reach a peak level (the first phase) at about 30 sec (237 ± 27%), then rapidly decreased to reach a steady level (the second phase) at about 10 min (28 ± 8%) (Fig. 1a). When the valvular strips were exposed to Ca-free PSS containing 2 mM EGTA, (Ca)i decreased gradually and reached a steady state after 5 min (–45 ± 6.2%) (Fig. 1b). The subsequent application of 10^{-6} M ET-1 induced a rapid rise in (Ca)i (peak level; 109 ± 14%) followed by a rapid decrease. The sustained phase of (Ca)i elevation seen in normal PSS was not observed in Ca-free PSS. In the presence of extracellular Ca, the application of 10^{-6} M ET-3 elevated (Ca)i (Fig. 1c). There was a first peak and a second sustained elevation. The peak level of the fluorescence ratio was 133 ± 17% and that of the sustained phase was 21 ± 2% at 10 min. In the absence of extracellular Ca, the peak level of the ratio induced by 10^{-6} M ET-3 was 68 ± 4% and there was no sustained phase (Fig. 1d). Both in the presence and absence of extracellular Ca, the responses of (Ca)i induced by ET-1 and ET-3 were concentration-dependent. In normal PSS, the maximum response obtained by ET-1 and ET-3 was 226 ± 8% and 136 ± 31%, respectively. The EC_{50} for ET-1 and ET-3 was 9.3 nM and 3.8 nM, respectively. In Ca-free PSS, the maximum response to ET-1 and ET-3 was 109 ± 14% and 68 ± 4%, respectively.

When the valvular strip was exposed to normal PSS containing 10^{-3} M Ni, an inorganic Ca entry blocker, the time course of changes in (Ca)i induced by 10^{-7} M ET-1 was nearly the same as that induced by 10^{-7} M ET-1 in Ca-free PSS containing 2×10^{-3} M EGTA. On the other hand, 10^{-5} M diltiazem, an organic Ca channel blocker, had no effects on the time course of changes in (Ca)i induced by ET-1 (submitted for publication). Thus, the first phase was composed of extracellular Ca-independent and -dependent components, while the second phase was exclusively extracellular Ca-dependent. The extracellular Ca-independent component of the first phase was due to the release of Ca from intracellular storage sites. The

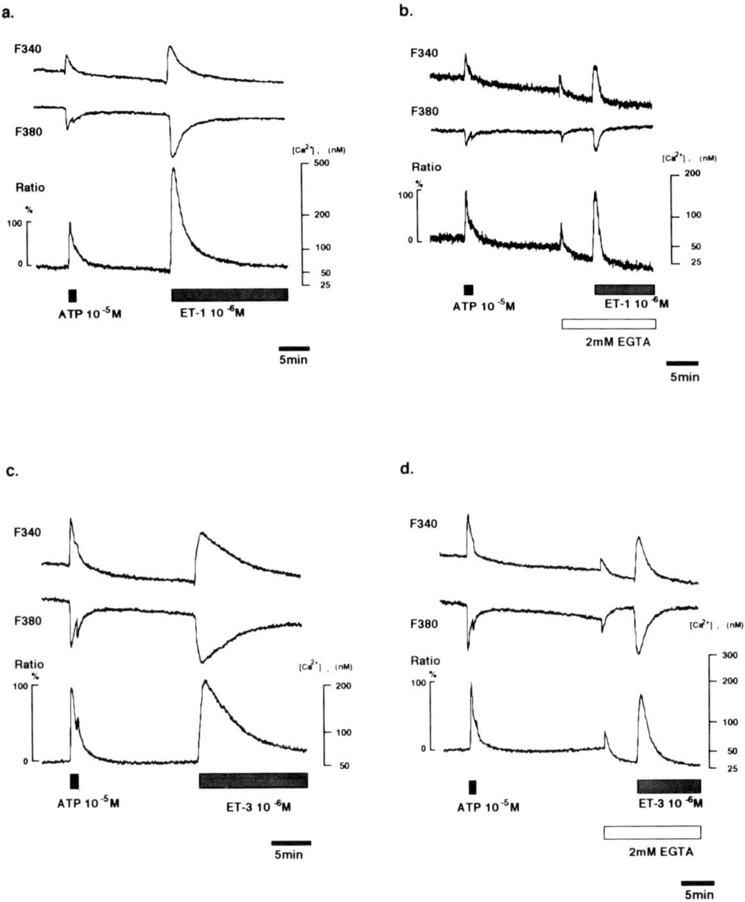

Fig. 1. Representative recordings of changes in endothelial (Ca)i induced by 10^{-6} M ET-1 in the presence (a) and absence (b) of extracellular Ca, and by 10^{-6} M ET-3 in the presence (c) and absence (d) of extracellular Ca. ET-1 and ET-3 caused rapid and transient elevations of (Ca)i both in normal PSS and Ca-free PSS. Only in normal PSS was the rapid elevation followed by a sustained elevation of (Ca)i [5].

second phase and part of the first phase of (Ca)i elevation were attributed to the influx of extracellular Ca.

Effect of IAP on ET-1-induced Ca transients in endothelial cells in situ *(submitted for publ.)*

In the presence of extracellular Ca, the levels of the first and the second phases of the elevations of (Ca)i induced by 10^{-7} M ET-1 were 221.7 ± 17.5% and 39.7 ± 4.9%, respectively. Pretreatment with IAP (300 ng/ml for 3h) markedly inhibited both the first and the second phases of elevation of (Ca)i induced by 10^{-7} M ET-1.

68

inhibited the elevation of (Ca)i and tension (Figs. 2a and 3). The steady-state levels of (Ca)i and tension development induced by U-46619 were 66.3 ± 2.0% and 92.3 ± 1.8%, respectively. The maximum reduction of (Ca)i and tension [(Ca)i = 53.9 ± 1.4%, tension = 70.3 ± 4.3%] were obtained with 0.6 nM ET-1. The duration of ET-1-induced decreases in (Ca)i and tension was 2–3 min. With a concentration of 3 nM or higher, there was no inhibition of (Ca)i or tension and ET-1 enhanced the elevations of (Ca)i and tension development induced by U-46619. When 10^{-6} M bradykinin was applied at the end of the specific protocols, the (Ca)i and tension decreased to nearly resting levels, thereby indicating that the endothelium was intact and functional. In the strips without the endothelium, ET-1 of low or high concentrations did not inhibit the elevations of (Ca)i and tension but did enhance the elevations of (Ca)i and tension induced by U-46619, in a concentration-dependent manner (Figs. 2b and 3). Thus, an inhibition of the contraction induced by low concentrations of ET-1 depends on the presence of an intact endothelium and the enhancing effects of high concentrations of ET-1 on U-46619-induced contractions are independent of endothelial cells.

To examine the contribution of prostacyclin and EDRF in the ET-1-induced relaxation, strips with an intact endothelium were pretreated with 10^{-5} M indomethacin or 10^{-4} M L-NNA. Decreases in (Ca)i and tension induced by ET-1 were slightly attenuated by indomethacin, but were significantly attenuated by L-NNA. Thus, while EDRF appears to play a major role in the ET-1-induced vasorelaxation, the minor contribution of prostacyclin in this relaxation cannot be neglected.

Discussion

In the previous study, using front-surface fluorometry and fura-2 loaded porcine coronary arterial strips without endothelium, we have found that ET-1 increases (Ca)i by means of both a release of Ca from the intracellular store and extracellular Ca-dependent mechanisms in smooth muscle cells [10]. Irrespective of the presence and the absence of extracellular Ca, the rapid increase in (Ca)i mainly relates to a release of Ca from the intracellular store, which practically determines the extent of tension development. The sustained phase of (Ca)i increase and tension development observed in the presence of extracellular Ca probably depends upon extracellular Ca. In coronary arterial smooth muscle cells, ET-1 induced a release of Ca from the intracellular store when the histamine-sensitive store was depleted, but not after depletion of the caffeine-sensitive store, suggesting that the ET-1-sensitive intracellular Ca store overlapped with the caffeine-sensitive one but not with the histamine-sensitive one. These ET-1-induced characteristic changes in Ca-transients and intracellular Ca store in porcine coronary arterial smooth muscle were essentially similar to those observed in the rat aortic vascular smooth muscle cells in primary culture [11,12]. In addition to Ca-dependent components of contraction, a Ca-independent component is involved in ET-1-induced contraction, since ET-1 evoked contraction accompanied with no apparent change in (Ca)i in strips in which the intracellular Ca stores had been depleted. The Ca-independent component of the

ET-1-induced contraction was markedly inhibited by the protein kinase C inhibitor, H-7, but not by the calmodulin inhibitor, W-7, suggesting that this Ca-independent component of contraction is probably mediated by a protein kinase C-related phosphorylation of contractile elements. EC_{50S} of ET-1-induced contractions in normal PSS and Ca-free PSS were 6×10^{-10} M and 9×10^{-9} M, respectively, and that of Ca-independent contraction was 2×10^{-8} M. Thus, although ET-1 may have multiple sites of action, it may induce contraction mainly due to Ca-dependent mechanism, with low EC_{50} values, in vascular smooth muscle cells [10].

This report directly shows that ET-1, the potent vasoconstrictor, also induces a decrease in (Ca)i and tension of vascular strips with intact endothelium. We showed evidence that ET-1 causes a transient relaxation (or inhibition of the contraction) of strips of the porcine coronary artery with an intact endothelium precontracted by U-46619. This relaxation is accompanied by a concomitant reduction in (Ca)i [6]. Since this effect disappeared with removal of the endothelium, the (Ca)i-decrease induced by ET-1 is probably mediated by actions through endothelial cells. The ET-1 induced endothelium-dependent relaxation was inhibited by L-NNA more effectively than by indomethacin, hence the ET-1-induced relaxation might be mediated mainly by EDRF in porcine coronary arteries.

It is generally accepted that the release of EDRF is regulated by a transient increase in (Ca)i of endothelial cells. We have provided evidence that ET-1 induces a greater extent of elevation of (Ca)i levels than does ET-3, both in the presence and absence of extracellular Ca, in the mature endothelial cells of porcine aortic valve leaflets. We have presented evidence for the involvement of an IAP sensitive G-protein in the signal transduction pathway activated by ET-1 in endothelial cells in situ (submitted for publ.). In normal PSS, ET-1 caused a rapid and transient elevation of (Ca)i (the first phase) followed by a sustained elevation (the second phase). Both in Ca-free PSS and in normal PSS containing 10^{-3} M Ni, the first phase was partially inhibited, while the second one was completely abolished. Thus, part of the first phase was independent of extracellular Ca and part of the first phase and all of the second phase were extracellular Ca dependent. An elevation of (Ca)i which is independent of extracellular Ca may be due to release of Ca from intracellular storage sites, presumably the endoplasmic reticulum, which was resistant to pretreatment with IAP. An increase in (Ca)i which depends on extracellular Ca may be due to the influx of Ca via plasma membrane and was almost completely inhibited by IAP. It is suggested that IAP-sensitive G-protein is involved in the ET-1-activated Ca influx mechanism in endothelial cells in situ. The G-protein that mediates intracellular Ca release, presumably via the phospholipase C/inositol trisphosphate cascade is IAP-insensitive.

In this report, we showed that part of elevation in (Ca)i induced by 10^{-7} M ET-1 was inhibited by low concentration of BQ-123 (10^{-7} M), a selective antagonist of ET_A receptor, while the rest was insensitive to higher concentration of BQ-123 (up to 10^{-5} M) (submitted for publ.). The extent of ET-1-induced elevation in (Ca)i resistant to BQ-123 was comparable to that of ET-3 (10^{-7} M)-induced one which was not affected by the ET_A antagonist. Thus, part of ET-1-induced Ca transients may be mediated by ET_A receptor, while Ca transient induced by ET-3 and BQ-123-

70

resistant component of Ca-transient induced by ET-1 may be mediated by some subtype other than ET_A, presumably ET_B. It is generally accepted that there is ET_B receptors in endothelial cells, however, the presence of ET_A receptors is a matter of controversy. In this report, in addition to the pharmacological evidence, it is shown that there is mRNA of ET_A receptors in the porcine aortic valvular endothelial cells in situ.

In conclusion, ET-1 is the only one of the ET-families made in the endothelial cells. ET-1 may play an important role in controlling vascular tonus not only by acting directly on vascular smooth muscle cells to induce contraction in a paracrine manner, but also by acting on endothelial cells to induce relaxation in an autocrine manner. ET_A receptors, as well as ET_B receptors, may play an important role when ET-1 acts on endothelial cells.

Acknowledgments

This study was supported in part by Grants-in-Aid for Developmental Scientific Research (No. 03557043) and for General Scientific Research (No. 04454268) from the Ministry of Education, Science and Culture, Japan.

References

1. Yanagisawa M, Kurihara H, Kimura S, Tomobe Y, Kobayashi M, Mitsui Y, Yazaki Y, Goto K, Masaki T. Nature 1988;332:411–415.
2. Inoue A, Yanagisawa Y, Kimura S, Kasuya Y, Miyauchi T, Goto K, Masaki T. Proc Natl Acad Sci 1989;86:2863–2867.
3. DeNucci G, Thomas R, D'Orleans-Juste P, Antunes E, Walder C, Warner TD, Vane JR. Proc Natl Acad Sci 1988;85:9797–9800.
4. Warner TD, DeNucci G, Vane JR. Eur J Pharmacol 1989;159:325–326.
5. Aoki H, Kobayashi S, Nishimura J, Yamamoto H, Kanaide H. Biochem Biophys Res Commun 1991;181:1352–1357.
6. Ushio-F M, Nishimura J, Aoki H, Kobayashi S, Kanaide H. Biochem Biophys Res Commun 1992;184:518–524.
7. Hirano K, Kanaide H, Abe S, Nakamura M. Br J Pharmacol 1990;101:273–280.
8. Grynkiewicz G, Poenie M, Tsien RY. J Biol Chem 1985;260:3440–3450.
9. Arai H, Hori S, Aramori I, Ohkubo H, Nakanishi S. Nature 1990;348:730–732.
10. Kodama M, Kanaide H, Abe S, Nakamura M. Biochem Biophys Res Commun 1989;174:228–235.
11. Miasiro N, Yamamoto H, Kanaide H, Nakamura M. Biochem Biophys Res Commun 1988;156:312–317.
12. Kai H, Kanaide H, Nakamura M. Biochem Biophys Res Commun 1989;158:235–243.

Endothelium-Derived Factors and Vascular Functions. T. Masaki, ed.

Endothelin receptor subtypes and mechanism of endothelin induced vasoconstriction

Tomoh Masaki, Aiji Sakamoto and Taijiro Enoki
Department of Pharmacology, Faculty of Medicine, Kyoto University, Kyoto 606, Japan

Introduction

Endothelin (ET) is a potent vasoconstrictive 21-amino acid peptide [1]. There are three endogenous isoforms with high homology, designated ET-1, ET-2 and ET-3 [2]. Vascular endothelial cell selectively produces and releases ET-1 to the basal side of the cell. The released ET-1 is thought to act on the underlying smooth muscle cells in paracrine manner and maintain the vascular tone [3].

On the other hand, there are at least two distinct but very similar ET-receptor subtypes designated ET_A and ET_B [4]. Vascular smooth muscle has ET_A receptor predominantly. Since specific ET_A-antagonist inhibits the ET-induced vasoconstriction, ET_A-receptor is thought to mediate vasoconstriction [4, 5]. ET_B receptor also exists in some vascular smooth muscle, for instance in rabbit saphenus vein, and mediates vasoconstriction [6]. However, in general, ET_B receptor preferentially exists on vascular endothelial cells and mediates release of relaxing factor, i.e. EDRF and prostacycline [4], although the regulatory mechanism of production of ET-1 and the relaxing factors has been yet to be elucidated. These results suggest that apparent functional difference between ET_A and ET_B receptors is ascribed to the difference in distribution of both subtypes of the receptor.

Then a question arose as to whether the true functional difference exists between ET_A and ET_B. Both subtypes belong to a superfamily of the seven transmembrane G-protein coupled receptor. Structurally, both are very similar but distinct. Human ET_A and ET_B receptors exhibit about 55% identity of the amino acid sequences [7].

In this article, first we describe the structural difference between both subtypes that is ascribed to the functional difference. Then, we will discuss the difference of the intracellular signal transduction systems stimulated by both subtypes. We also discuss important signaling pathways in ET-induced vasoconstriction.

Structure-function relationship of ET-receptor

ET_A receptor has a high affinity to ET-1 and ET-2, but a low affinity to ET-3, while ET_B receptor has an equipotent affinity to all of the three isopeptides of ET.

To inquire into the location of the subtype-specific determinants within the ET-receptor, we constructed a series of recombinant chimera by progressing substitution of the ET_A with the ET_B cDNA, and expressed them in a heterologous cell line, COS-7 cell. Then we characterized those receptors in terms of ligand selectivity by competitive radioligand binding assays [8]. The cell was reacted to ^{125}I-ET-1. The bound radioligand was displaced in a competitive manner by ET-1, ET-3 or a specific ET_A-antagonist BQ-123. The apparent Ki value of ET-3 to ET_A, ET_B or various chimeric receptor was divided by the Ki value of ET-1 to the corresponding receptor. The resulting value R_{ET-3} shows the relative affinity of each receptor to ET-3 against ET-1 (Table 1). The relative affinity of each receptor to BQ-123 against ET-1, R_{BQ123}, is also shown in Table 1. The result shows that the structure of the ET_B receptor spanning from fourth transmembrane domain (TMD IV) to TMD VI including adjoining loops is important in high affinity binding of ET_B receptor to ET-3. It also demonstrated that the ET_A sequences from TMD I through III and VII with adjacent loops are necessary and sufficient for high affinity to BQ-123. Further detailed experiments demonstrated that BQ-123 binds to the first extracellular loop of the ET_A receptor [9].

A chimeric receptor that has the TMDs IV-VI and adjacent loop regions from ET_B inserted into the remaining regions from ET_A exhibited high affinities to both BQ-123 and ET-3, which exerted antagonistic and agonistic activities, respectively [8]. This chimera receptor exhibited specific binding to the N-terminally truncated ET_B agonists BQ-3020 and IRL-1620 with similar level to wild type ET_B. It is also noted that all of the endogenous ET as well as the ET_B-agonists mentioned above, have the same amino acid sequence at the C-terminal side despite of the distinct sequences at N-terminal side of the ET_B. The C-terminal sequence of ET agonists is essential to the binding to both types of the receptor. Since BQ-123 inhibits the binding of those ligands to the receptor, the C-terminal sequence binds to the TMD I-III plus VII and adjacent domains. ET-agonists may induce activation of intracellular signal transduction system through this binding.

In contrast, structure of the N-terminal side of the ET_B is important in affinity difference of the agonist-receptor binding. The TMDs of ET_B from IV through VI do not require any loop structures of ET_B for their high affinity. In contrast, ET_A receptor requires the interaction of the corresponding subdomain of ET_A and the N-terminal loop portion of ET-1. The N-terminal loop portion of ET-1 and the TMDs from IV through VI with the adjoining loop regions of ET_A receptor are involved in the selectivity of the ligands [8].

These results suggest that ET-receptor consists of two functional domains, i.e. ligand-binding and ligand-selection domains. ET_A and ET_B are functionally different from each other in their ligand-selection domains.

Table 1. Relative affinity of ET_A, ET_B and various chimera receptor to ET-3 and BQ-123 against ET-1

Receptor construct	$R_{ET\text{-}3}$	R_{BQ123}
Wild type B	2.1	33,000
A(N)B(I-C)	5.1	18,000
A(N-I)B(II-C)	1.7	27,000
A(N-II)B(III-C)	5.8	11,000
A(N-III)B(IV-C)	1.2	120
A(N-IV)B(V-C)	38	210
A(N-V)B(VI-C)	190	130
A(N-VI)B(VII-C)	450	79
A(N-VII)B(C)	610	6.3
Wild-type A	290	7.1
B(N)A(I-C)	280	19
B(N-I)A(II-C)	120	85
B(N-II)A(III-C)	140	6,100
B(N-III)A(IV-C)	420	20,000
B(N-IV)A(V-C)	390	16,000
B(N-V)A(VI-C)	55	37,000
B(N-VI)A(VII-C)	6.0	40,000
B(N-VII)A(C)	3.4	16,000

N, I-VII, and C designate the amino-terminal extracellular tail, transmembrane helices I-VII, and carboxyl-terminal cytoplasmic tail respectively, plus the adjacent loop regions. For example, A(N-I)B(II-C) means the chimeric receptor that contains ET_A sequence from N-terminus to first transmembrane domain and ET_B sequence from second transmembrane domain to C-terminus. (Table I of reference 8 was modified.)

Intracellular signal transduction systems stimulated by ET_B

Numerous reports have demonstrated that ET_B stimulated phospholipase C, D, A2, protein kinase C and Na^+/H^+ exchanger as well as several ionic channels [4,10]. Those responses are mediated by both ET_A and ET_B receptors. Difference between ET_A and ET_B-mediated responses will be clear, if the responses induced by both types of the ET-receptor are compared under the same condition. Nakanishi's group reported the ET-1-stimulated intracellular signal transductions that was mediated by ET_A or ET_B expressed on CHO cells [11]. Both receptors showed a rapid and marked stimulation of phosphatidylinositol hydrolysis and arachidonic acid release in response to agonist interaction. Only one difference between ET_A and ET_B was the activity of adenylate cyclase. ET_A mediated the accumulation of cyclic AMP formation, whereas ET_B displayed an inhibitory action on the forskolin-stimulated cyclic AMP accumulation. However, the ET_B-mediated inhibition of cAMP accumulation was partial. The ET_A-mediated cAMP accumulation was also partially inhibited by pertussis toxin. Those results suggest that several members of G-protein are coupled in those ET-receptor systems.

74

In isolated vascular beds, ET elicits slow developing and long-lasting vasoconstriction. On the other hand, stimulation of ET-receptor elicits a rapid rise in cytosolic free calcium ion concentration followed by a sustained increase in cytosolic free calcium ion in target cells including smooth muscle cells [10,12]. The increase in cytosolic free calcium ion induces contraction of smooth muscle. Since the ET-induced vasoconstriction is slow developing contraction, the later sustained increase in cytosolic free calcium ion appears to be important in ET-induced vasoconstriction. It is generally accepted that the transient increase in cytosolic free calcium ion is ascribed to the calcium ion released from intracellular calcium pool by inositoltrisphosphate [10]. The later sustained increase in cytosolic free calcium ion is ascribed to the exterior calcium ion.

In Rat-1 cells, ET-1 significantly increased calcium influx at a low concentration of ET-1, with peak stimulation occurring at about 2×10^{-11} M [13]. In this case the calcium influx was inhibited by elevation of the intracellular calcium concentration. On the other hand, at such low concentrations of ET-1, production of inositol trisphosphate and diacrylglycerol were low, consequently mobilization of intracellular free calcium ion rarely occurred. EC_{50} value of production of inositolphosphate stimulated by ET-1 was around 10^{-9} M. Since physiological concentration of ET-1 in vascular wall is estimated to be very low, ET-1 may not elicit activation of phospholipase C in vascular beds in physiological circumstance. Recent experiments on binding of ET-1 to ET_A-receptor demonstrated that association of ET-1 to the ET_A receptor was very slow [14]. Steady state was reached usually in 60 minutes after administration of 2×10^{-11} M ET-1. Therefore, only a few percent of the amount of ET-1 can bind to the receptor whithin 30 seconds in which calcium transient occurs at this concentration of ET-1. However, the higher the concentration of ET-1, the more rapidly association of ET-1 with the receptor and consequently accumulation of inositolphosphate occur. Therefore, these results suggest that the calcium transient is not important in vasoconstriction under physiological conditions.

Then, the next question arises as to what kind of mechanism is involved in ET-induced calcium influx at the later sustained increase in cytosolic free calcium ion. In an isolated smooth muscle of porcine coronary artery, the later sustained phase in cytosolic free calcium ion was inhibited by a dihydropyridine calcium antagonist. [15]. The whole-cell patch-clamp technique, using the smooth muscle cells isolated from porcine coronary artery, demonstrated that ET-1 increased the peak calcium channel current. This increased current was almost completely inhibited by the addition of 1 μM nicardipine [15]. The dose-response curve of the ET-1-induced vasoconstriction of porcine coronary artery was shifted to the right by nicardipine, suggesting that voltage-operated calcium channel was involved in this mechanism [16]. However, a number of papers demonstrated that the inhibition of ET-1-induced contraction by dihydropyridine calcium antagonist was only partial in many other vascular beds [17]. Also, the later sustained elevation of cytosolic free calcium ion was not always abolished by dihydropyridine calcium antagonist. For instance, in A-10 cells, nickel ion but not nicardipine promptly negated the second plateau phase, suggesting that other type of cation channel is involved in this mechanism [12].

To elucidate this mechanism, rat fibroblast L-cell was transfected with human ET_A cDNA [8]. The expressed ET_A receptor was stimulated by a low concentration of ET-1, $(10^{-10}$ M). As observed in other cases, the transient elevation of cytosolic free calcium ion followed by a sustained phase was observed. Removal of calcium ion from outside medium completely inhibited both the transient and the sustained phase. Dihydropyridine calcium antagonist, nicardipine, had no effect on the ET-induced cytosolic free-calcium ion in this case. We demonstrated that ET-1 stimulated the influx of calcium ion through non-selective cation channel, and that mefenamic acid, a specific inhibitor of non-selective cation channel, inhibited the calcium influx. These results suggest that both voltage dependent calcium channel and non-selective cation channel mediated ET-1-induced vasoconstriction. Relative importance of either of the two channels depends on the vascular bed employed. Indeed, in rabbit aorta, vasoconstriction induced by a low concentration of ET-1 was inhibited either by mefenamic acid or by dihydropyridine calcium antagonist (unpubl. data).

ET-1-induced sustained contraction in porcine coronary artery was also observed without an increase in cytosolic free calcium ion level. This contraction was attenuated by administration of staurosporine or H-7, suggesting an activation of protein kinase C by ET-1 and consequently an increase in the calcium sensitivity of the contractile proteins of the smooth muscle [16, 18]. The activation of protein kinase C by ET-1 can be ascribed to the increase in the amount of diacylglycerol. It was demonstrated that ET-1 stimulated diacylglycerol (DG) formation in biphasic manner, i.e. an early rapid phase peaking at 30 sec followed by a sustained phase lasting for more than 10 minutes [19]. Since the ET-1-induced activation of phospholipase C (PLC) occurs in a short period [12], the initial peak of the DG formation corresponds to the PLC activation. ET-1 was shown to activate also phospholipase D (PLD) and produce DG [20]. After the stimulation of rat aorta by ET-1, accumulation of phosphatidylethanol and phosphatidic acid occurred in time-dependent and sustained manner [20]. Since DG is produced from phosphatidic acid, the ET-induced sustained contraction may be due to the activation of PLD. However, activation of PLD by ET-1 occurred only at high concentration of ET-1. Therefore, those pathways including PLC and PLD may be important in pharmaco-logical constriction induced by exogenous ET-1, but not in ET-induced constriction in physiological condition.

Conclusion

There are at least two subtypes of ET-receptor, designated ET_A and ET_B. In vascular beds, ET_A exists on smooth muscle and mediates vasoconstriction. ET_B exists on endothelial cells and mediates production and release of relaxing factors. Thus, apparent functional difference between both subtypes depends on the tissue distribution of both subtypes. Analysis of structure and function relationship of both subtypes demonstrated that affinity difference of both subtypes for endogenous ligands is ascribed to difference of the structure of transmembrane domains IV-VI including adjoining loops. This domain determines the high affinity of the ET_B receptor for ET-3.

Activated ET-receptor stimulates many intracellular signal transduction systems and ion channels, suggesting that both subtypes of ET-receptor stimulate the second messenger systems via several types of G-protein. Only one difference between both subtypes in their stimulation of intracellular signaling systems is the accumulation of cAMP.

Regarding ET-induced vasoconstriction, activation of non-selective cation channel is important in physiological condition, because it is activated at low concentration of ET-1. L-type calcium channel is also important in ET-induced vasoconstriction in several types of vascular beds. However, activation of PLC or PLD is important only in pharmacologically induced vasoconstriction by exogenous ET_B.

References

1. Yanagisawa M, Kurihara H, Kimura S, Tomobe Y, Kobayashi M, Mitsui Y, Yazaki Y, Goto K, Masaki T. Nature 1988;332:411–415.
2. Inoue A, Yanagisawa M, Kimura S, Kasuya Y, Miyauchi T, Goto K, Masaki T. Proc Natl Acad Sci USA 1989;86:2863–2867.
3. Masaki T. Annal NY Acad Sci (in press).
4. Sakurai T, Yanagisawa M, Masaki. Trends Pharmacol Sci, 1992;13:103–108.
5. Ihara M, Noguchi K, Saeki T, Fukuroda T, Tsuchida S, Kimura SL, Fukami T, Ishikawa K, Nishikibe M, Yano M. Life Science 1991;50:247–255.
6. Cristol JP, Warner TD, Thiemermann C, Vane JR. Br J Pharmacol 1993;108:776–779.
7. Sakamoto A, Yanagisawa M, Sakurai T, Takuwa Y, Yanagisawa H, Masaki T. Biochem Biophys Res Commun 1991;178:656–663.
8. Sakamoto A, Yanagisawa M, Sawamura T, Enoki T, Ohtani T, Sakurai T, Nakao K, Toyo-oka T, Masaki T. J Biol Chem 1993;268:8547–8553.
9. Adachi M, Hashido K, Trzeciak A, Watanabe T, Furuichi Y, Miyamoto C. J Cardiovasc Pharmacol (in press).
10. Masaki T. Endocrine Rev 1993;14:256–268.
11. Aramori I, Nakanishi S. J Biol Chem 1992;267:12468–12474.
12. Takuwa Y, Kasuya Y, Takuwa N, Kudo M, Yanagisawa M, Goto K, Masaki T, Yamashita K. J Clin Invest 1990;80:653–658.
13. Muldoon LL, Endlen H, Rodland KD, Magun BE. Am J Physiol 1992;260:C1273–C1281.
14. Waggoner WG, Genova SL, Rash VA. Life Sci 1992;51:1869–1876.
15. Goto K, Kasuya Y, Matsuki N, Takuwa Y, Kurihara H, Ishikawa T, Kimura S, Yanagisawa M, Masaki T. Proc Natl Acad Sci USA 1989;86:3915–3918.
16. Kasuya Y, Ishikawa T, Yanagisawa M, Kimura S, Goto K, Masaki T. Am J Physiol 1989;257:H1828–H1835.
17. Ohlstein EH, Horohonich S, Hay DWP. J. Pharmacol Exp Ther 1989;250:548–555.
18. Kodama M, Kanaide H, Abe S, Hirano K, Kai H, Nakamura M. Biochem Biophys Res Commun 1989;160:1302–1308.
19. Sunako M, Kamahara Y, Hirata K, Tsuda T, Yokoyama M, Fukuzaki H, Takai Y. Hypertension 1990;15:84–88.
20. Liu Y, Geisbuhler B, Jones AW. Am J Physiol 1992;262:C941–949.

Endothelium-Derived Factors and Vascular Functions. T. Masaki, ed.

Endothelin and renal tubules

Shunya Uchida, Fumi Takemoto, Koji Yoshitomi and Kiyoshi Kurokawa
The First Department of Internal Medicine, University of Tokyo Faculty of Medicine, Showa General Hospital, 2-450 Tenjin-Cho, Kodaira, Tokyo 187; The Second Department of Internal Medicine, Kyushu University Faculty of Medicine, Fukuoka, Japan

Summary

Endothelin-1 (ET-1) may play a pivotal role in regulating renal function. In addition to regulate renal hemodynamics ET-1 may act directly on renal tubules. To examine which tubule segment has ET-1 binding capacity, microdissected rat tubules underwent $[^{125}I]$ET-1 binding studies. Predominant ET-1 binding was found in collecting tubules, indicating that target renal tubules for ET-1 are collecting ducts. Since BQ3020 (specific ET_B agonist) but not BQ123 (specific ET_A antagonist) inhibited the ET-1 binding in cortical collecting ducts (CCD), ET receptor subtype in CCD was concluded to be ET_B subtype. The specific ET-1 binding to CCD was desensitized by 20 min pretreatment with arginine vasopressin (AVP). This ET_B receptor down-regulation occurred dose-dependently and was mimicked by forskolin and dibutyryl cAMP, suggesting the mechanism being via a cAMP-mediated pathway. By contrast, ET_A expressed in aortic smooth muscle cells did not show cAMP-dependent down-regulation. Finally we conducted electrophysiological experiments using rabbit CCD perfused in vitro where ET-1 decreased lumen negative transepithelial voltage in association with an increase in intracellular Ca^{2+}. In conjunction with an increase in total cellular membrane resistance by ET-1, it is suggested that ET-1 may inhibit Na^+ entry from the luminal side of CCD. This conclusion well explains the inhibition of the tubular Na^+ reabsorption observed in vivo clearance experiments of rats. Our recent attempt by RT-PCR using microdissected renal tubules revealed collecting ducts may also be production sites for ET-1. Taken all together ET-1 may function as an autocrine/paracrine factor in the collecting duct of the kidney and thus may be involved in certain pathological derangements of the kidney.

Introduction

Endothelin (ET) consists of three isopeptides (ET-1, ET-2 and ET-3), each of which displays subtle differences in potency for constricting vascular smooth muscle cells [1] and in other biological effects against target tissues. These dissimilarities maybe derive from the difference in binding capacity of the three isopeptides to their target tissues. Indeed the cDNAs for two ET receptor subtypes, named ET_A and ET_B, were cloned [2,3].

Our early study using in vivo rat perfusion demonstrated that ET-1 dramatically decreased urine volume via decrease in renal plasma flow and glomerular filtration rate (GFR) [4]. The effect was due to strong vasoconstricting potential of ET-1 as seen in other tissue. What surprised us was that ET-1 increased fractional excretion of Na^+ in spite of lowered GFR, indicating that ET-1 may have certain natriuretic potential unlike other vasoconstricting peptides such as angiotensin II.

We also found that ET-1 caused a biphasic increase in intracellular ionized

78

calcium concentration ($[Ca^{2+}]i$) which consists of an initial rapid rise followed by a second more sustained elevation in $[Ca^{2+}]i$ in the collecting ducts [5]. In perfused CCD Ca^{2+} removal from the superfusate resulted in attenuation of the second phase of $[Ca^{2+}]i$ but not the first spike. These data suggest that the initial peak of the ET-1-evoked rise in $[Ca^{2+}]i$ comes largely from intracellular Ca^{2+} store and that the second sustained rise in $[Ca^{2+}]i$ is largely due to increased Ca^{2+} influx [5].

However, little is known about the function of ET-2 and ET-3 in the kidney in part because of the lack of information on the intrarenal distribution of ET receptor subtypes. To further understand the biological roles of ET isopeptides in the kidney, the present study was undertaken to clarify the distribution of ET-1 and ET-3 binding along the rat nephron and to determine receptor subtypes based on their binding properties. Moreover, we examined effects of ET-1 on electrophysiological parameters and $[Ca^{2+}]i$ to elucidate the cellular mechanism of natriuretic action of ET-1 using rabbit CCD perfused in vitro.

Methods

Tubule microdissection. Individual glomeruli and different nephron segments were dissected freehand under stereomicroscopic observation in modified Hanks' solution at 4°C. Parts of microdissected nephron were: glomeruli; the early, middle and terminal portions of the proximal tubule (Sl, S2 and S3, respectively); cortical and medullary thick ascending limbs of Henle's loop (CAL and MAL, respectively); the distal convoluted tubule (DCT); and the cortical, outer medullary and inner medullary collecting ducts (CCD, OMCD and IMCD, respectively). Tubule segments (total length 2–5 mm) or 8–10 glomeruli were transferred onto an 8 × 8 mm piece of aluminum foil placed on ice.

Receptor binding assay. [^{125}I]ET-1, [^{125}I]ET-3 (specific activity of each $\approx$ 2000 Ci/mmol, Amersham, Denton, MD), and respective homologous unlabeled peptides (Peptide Institute, Osaka, Japan) were used for radioreceptor assays. The binding solution consisted of Eagle's modified essential medium (MEM) (12 µl) containing HEPES 20 mM, $NaHCO_3$ 0.03% and BSA 0.1%, pH 7.4. After incubation at 4°C for 4 h, radioactivity bound to the nephron segments was counted in a gamma counter (Aloka, Tokyo, Japan) for 20 min.

Binding capacities of individual nephron segments to ET-1 or ET-3 were obtained using 250 pM [^{125}I]ET-1 or [^{125}I]ET-3. ET binding capacities to isolated tubules segments and glomeruli were normalized per unit segment length and per single glomerulus, respectively. Then, saturation binding and competition binding of ET-1 or ET-3 in CCD as representative tubules were conducted. Maximal binding capacity (Bmax) and binding affinity (Kd) of the ET receptor in CCD were calculated by Scatchard analysis of the obtained data. Details of these experiments are described elsewhere [6].

ET receptor subtype identification. In order to identify ET receptor subtype in CCD or aorta smooth muscle cells we used a specific ET_A antagonist, BQ-123 [7], and a specific ET_B agonist, BQ-3020 [8] (both kindly provided by Banyu Pharmaceutical Co., Tsukuba, Japan). [^{125}I]ET-1 binding in the presence of these chemicals and unlabeled ET-1 were compared and the difference between [^{125}I]ET-1 binding with and without unlabeled ET-1 were expressed as 100%.

In vitro microperfusion of CCD and cell Ca^{2+} measurement. CCD was perfused in vitro according to the method of Burg et al. [9] with slight modifications as reported previously from our laboratory [10]. The bathing solution temperature was controlled at 37°C by adjusting the temperature of the heating jacket. The flow rate of the bathing solution was adjusted at approximately 6–9 ml/min, allowing the bath fluid exchange within 1 sec.

The transepithelial voltage (VT) was measured through a perfusion pipette, which was connected to one channel of a dual electrometer (Duo 773, WP-Instruments, New Haven, CT). A flowing boundary 1 mol/l KCl electrode, which was connected to a calomel half cell, was placed at the outflow of the bath and served as a system ground.

Intracellular Ca^{2+} was measured by using a fluorescence microscope photometry system (OSP-3, Olympus Co., Tokyo). Isolated microperfused rabbit CCD were incubated in 200 μl standard bicarbonate buffer solution containing 10 mM acetoxymethyl ester of fura 2 (fura 2 AM, Dojin Biochemical, Kumamoto, Japan) at 37°C for 30 min. Tubular cells were alternately excited at 340 nm and 380 nm and the emission was monitored at 510 nm. [Ca^{2+}]i was calculated based on the formula described by Grynkiewicz et al. [11]. Details of the electrophysiological experiments are described elsewhere [12].

PCR amplification of ET-1 mRNA in microdissected tubules. To detect whether microdissected fresh renal tubules contain messenger RNA for ET-1 we utilized reverse transcription-polymerase chain reaction (RT-PCR) method. RT was performed with oligo-dT primers using a cDNA synthesis kit (Boehringer Mannheim, Mannheim, Germany). PCR was performed using the GeneAmp DNA amplification reagent kit (Perkin-Elmer Cetus, Norwalk, CT). Amplified PCR products were subjected to Southern analysis to verify their authenticity. The details of this procedure are described elsewhere [13].

Statistics. All values are expressed as means ± SEM. When comparing three or more groups, such as for the distribution of ET-1 and ET-3 binding along the nephron, statistical analysis was performed using one way ANOVA. Whenever ANOVA indicated significance, comparisons between any two groups were made with the non-paired Student's *t*-test. A p value of less than 0.05 was considered significant.

membrane, effects of a Ca^{2+} channel blocker, nicardipine, was tested. 10^{-8}–10^{-5} M nicardipine in the bath decreased $[Ca^{2+}]i$ in a dose-dependent manner, suggesting that dihydropyridine-sensitive Ca^{2+} channels may exist in the basolateral membrane. Moreover, at the dose of 10^{-7} M nicardipine, the effect of ET-1 on $[Ca^{2+}]i$ disappeared, whereas $[Ca^{2+}]i$ response to AVP was still observed, indicating that ET-1 causes Ca^{2+} influx from the basolateral membrane via dihydropyridine-sensitive Ca^{2+} channels [5]. Furthermore, the effects of ET-1 on V_T, V_B and Ri were abolished in the presence of Ca^{2+} channel blocker, indicating that the Ca^{2+} entry from the basolateral membrane via Ca^{2+} channels is essential for the electrophysiological response to ET-1. Thus the initial depolarization of the basolateral membrane together with an increase in $[Ca^{2+}]i$ may be explained by Ca^{2+} influx from the basolateral membrane.

Since the late hyperpolarization of the luminal membrane was associated with an increase in membrane resistance, it is likely that ET-1 inhibits Na^+ conductance in the luminal membrane. To confirm this hypothesis, the effect of ET-1 was tested in the presence of Na^+ channel blocker, amiloride. When 10^{-5} M amiloride was added to the lumen, the luminal membrane was hyperpolarized from 71.0 ± 4.7 mV to 79.7 ± 3.8 mV, and Ri increased from 231.4 ± 23.4 MΩ to 267.1 ± 26.3 MΩ. In this setting, electrical parameters did not change in response to ET-1, indicating that Na^+ channel plays a pivotal role in the action of ET-1.

Detection of ET-1 mRNA in rat nephron by RT-PCR method

ET-1 mRNA was only detected in cortical, outer medullary, and inner medullary collecting ducts in addition to isolated fresh glomeruli. Abundance of ET-1 mRNA in other nephron segments such as proximal nephron and thick ascending limbs were virtualy negligible at the basal state. Thus, the sites for ET-1 production in the nephron are identical to those for ET-1 action.

Discussion

In the present study, we examined distribution profiles of ET-1 and ET-3 binding sites along the rat nephron. ET-1 binding was located primarily along the collecting duct, with the highest capacity in IMCD, while binding to CCD or OMCD was about half that to IMCD. In contrast, binding to S2, S3, CAL, MAL and DCT was barely detectable under our experimental conditions.

The major site of ET-1 action in renal tubules has been ascribed to the collecting duct since the initial study by Zeidel et al. who demonstrated inhibition of Na^+/K^+-ATPase activity by ET-1 in rabbit IMCD cells [14]. Later Tomita et al. reported an inhibitory effect of ET-1 on AVP-stimulated cAMP accumulation in rat CCD, OMCD and IMCD [15]. We have recently demonstrated that ET-1 increases intracellular Ca^{2+} concentration in mouse collecting ducts dissected from the cortex to the inner medulla [5]. The present results therefore agree quite well with the current notion that ET-1 principally acts on cortical and medullary collecting ducts via binding to its putative receptors.

As for ET-3 binding along the nephron ET-3 clearly bound not only to glomeruli but also to tubule segments, such as Sl and the collecting duct. The overall profile of ET-3 binding along the rat nephron was similar to that of ET-1 binding. Moreover, ET-1 and ET-3 bindings to collecting tubule segments were virtually equal in quantity, implying that each ligand may bind to the same class of ET receptor with a comparable affinity.

We therefore examined which ET receptor subtype was expressed in the CCD segments. Scatchard analyses of competitive binding studies using homologous ET isopeptide in CCD were performed. ET-1 binding to CCD showed a single class of receptor, as did ET-3 binding. The Kd value for ET-1 was almost identical to that for ET-3. Moreover, ET-1 binding to CCD was commensurably displaced by addition of unlabeled ET-3, indicating that ET-1 and ET-3 are able to bind to the same receptor subtype with an equivalent affinity. These results strongly suggest that the ET receptor in the CCD is the ET_B subtype. We further utilized a specific ET_A antagonist and ET_B agonist to confirm this interpretation. Only the ET_B agonist (BQ-3020) inhibited $[^{125}I]ET$-1 binding to the same degree as unlabeled ET-1. This result provides further evidence of the predominant expression of the ET_B receptor subtype in CCD.

The cDNAs for ET_A and ET_B receptors have been cloned and their mRNAs are expressed in the whole kidney [2,3]. Recently, a precise localization of each receptor mRNA has been elucidated in the rat nephron using RT-PCR by Terada et al. [16]. They found that ET_A receptor mRNA was detected only in glomeruli and the vascular system, whereas ET_B receptor mRNA was expressed in cortical and medullary collecting ducts in addition to glomeruli. Our data for CCD are compatible with the predominant expression of ET_B receptor mRNA. More recently, however, the presence of ET_A receptors in cultured as well as freshly isolated IMCD cells has been demonstrated by Kohan et al. [17]. Both ET_A and ET_B may exist at least in IMCD.

Comparison of amino acid sequences of ET_A and ET_B receptors indicated a delicate difference in terms of possible phosphorylation sites: ET_B contains an unique site to be phosphorylated by PKA while ET_A lacks such a motif. This motif consists of Arg-Arg-Lys-Ser, agreeing with the consensus sequence of Arg/Lys-Arg/Lys-X-Ser/Thr as the most efficient substrate for PKA, and locates in the third cytoplasmic loop bridging fifth and sixth transmembrane regions of the ET_B. By contrast, the corresponding third cytoplasmic loop of ET_A receptor subtype is likely to conserve two consensus sequences for PKC-dependent phosphorylation: Ser-Leu-Arg and Ser-Glu-His-Leu-Lys, both matching the motif of $SX_{0-3}R$. Thus we hypothesized the phosphorylation by PKA at the unique corresponding site found only in ET_B may change ligand-binding characteristics, hence allowing modulation of ET_B-mediated functions of ET isopeptides by PKA activation.

In the present study, we first demonstrated that physiological concentrations of AVP inhibit ET-1 binding to CCD, the phenomenon called "heterologous desensitization" because a reduction in ET-1 binding capacity is caused by

84

heterologous ligand. As low a concentration as 10^{-11} M of AVP significantly downregulated ET-1 binding. Another interesting feature is that this desensitization occurred as early as 20 min after AVP exposure. The inhibitory effect of AVP on ET-1 binding to CCD appears to be due to cAMP-PKA axis, because forskolin and DBcAMP faithfully reproduced its effect. It was further confirmed by the fact that H-89, the specific PKA inhibitor, blunted AVP-induced desensitization of ET-1 binding. Thus the phosphorylation by PKA may play a central role in desensitization of ET-1 receptor.

On the other hand, in the preparation of endothelium-denuded aortic strips which predominantly express ET_A subtype, neither forskolin nor DBcAMP caused desensitization of ET-1 binding, suggesting that ET receptor desensitization by PKA is quite specific for ET_B receptor subtype. Although vascular smooth muscle cells possess a machinery to increase intracellular cAMP content, ligands capable of stimulating cAMP generation may not influence ET-1 binding to ET_A probably due to a lack of PKA-dependent phosphorylation motif. These results strongly suggest that ET_B receptor subtype is specifically conferred a flexibility to be regulated by cAMP-dependent mechanism unlike ET_A receptor subtype.

What is the possible mechanism of ET receptor desensitization we observed in the present study? Desensitization of ET_B receptor in response to AVP can be characterized in three aspects. First, the desensitization occurs very rapidly; second, it occurs at physiological concentrations of AVP; third, the desensitization is due to a reduction in binding affinity but not to a reduction in maximal binding capacity. Studies on the mechanisms of β-receptor desensitization have provided valuable insights into the molecular events of receptor desensitization. Types of receptor desensitization have now been classified into four major mechanisms: sequestration, degradation and inactivation of the receptor, G protein inactivation, and transcriptional control. Possibilities of both sequestration of the receptor and transcriptional control of de novo synthesis are unlikely because the former takes place at micro molar concentration of the ligand and the latter takes a prolonged time of exposure to the ligand such as several hours. Moreover, both mechanisms as well as receptor degradation will induce a reduction of maximal binding capacity without affecting binding affinity. By contrast, short-term desensitization, either homologous or heterologous, may underlie receptor phosphorylation and an impairment of G protein coupling.

Therefore, it may be that AVP increases cAMP formation, activates PKA, then phosphorylates G protein binding domain of ET_B receptor, leading to uncoupling of heterotrimeric G protein from ET_B receptor, which finally causes rapid desensitization of the ET_B receptor. The results further suggest the interrelation or mutual inhibition between AVP and ET-1 in the collecting duct.

The experiment using in vitro perfused rabbit CCD demonstrated that ET-1 inhibits Na^+ channel in the apical membrane of CCD. This notion is supported by the following observations [12]. First, ET-1 hyperpolarized the apical membrane concomitant with an increase in membrane resistance, suggesting that ET-1 decreases the apical membrane Na^+ conductance. The apical membrane conductance of the CCD cell is mainly determined by Na^+ and K^+ conductances. If ET-1 inhibits

K^+ conductance, it should depolarize, rather than hyperpolarize, the apical membrane. Second, amiloride added to the lumen completely blocked the electrophysiological responses to ET-1, results consistent with a notion that intactness of the amiloride-sensitive Na^+ channel is essential for the change in apical membrane voltage caused by ET-1.

It has been reported that the intracellular Ca^{2+} signaling pathway plays an important role in the physiological action of ET-1. We found that an addition of nicardipine or an elimination of Ca^{2+} in the bathing fluid resulted in no response of $[Ca^{2+}]i$ to ET-1. The lack of the Ca^{2+} response to ET-1 in these conditions was not due to an inability of the cell to respond to hormonal stimuli since the Ca^{2+} response to AVP was clearly detectable. These data suggest that the major action of ET-1 in increasing $[Ca^{2+}]i$ is due to $[Ca^{2+}]i$ entry through a pathway sensitive to dihydropyridine [5].

The present study also demonstrated that a rise in Ca^{2+} through the basolateral membrane evoked by ET-1 is essential for the inhibition of the luminal Na^+ conductance because the electrophysiological responses to ET-1 were completely abolished in the presence of nicardipine. It is unlikely, however, that the change in Ca^{2+} per se is sufficient to inhibit luminal Na^+ conductance since decreases in Ca^{2+} in response to Ca^{2+} removal or to nicardipine addition to the bath solution did not affect V_T. Thus, a rise in Ca^{2+} together with other cellular events brought about by ET-1, e.g. protein kinase C activation, leads to inhibition of luminal Na^+ conductance. In any events, we propose that binding of ET-1 to the basolateral membrane of CCD stimulates dihydropyridine-sensitive Ca^{2+} channel, leading to an increase in $[Ca^{2+}]$ together with other cellular events, which in turn inhibits Na^+ channel in the luminal membrane, and that such effects of ET on CCD may explain, at least in part, the lack of a fall in FE_{Na} despite decreases in RPF and GFR by infusion of ET-1 to the kidney.

An increasing body of evidence indicates that ET-1 is synthesized in tubular epithelial cells as well as in glomerular endothelial, epithelial cells and mesangial cells [18,19]. Our recent attempt to clarify ET-1 mRNA localization, using the RT-PCR method, disclosed that ET-1 mRNA is expressed mainly in cortical and medullary collecting ducts in addition to the glomeruli [13]. These sites of ET-1 synthesis appear almost identical to the pattern of ET-1 binding along the nephron elucidated by our present study. It is therefore possible that ET-1 may function in the renal tubules as an autocrine or paracrine factor and may play an important role in certain pathological conditions of the kidney such as ischemic acute renal failure.

Acknowledgments

We thank Drs. M. Ihara and M. Yano (Banyu Pharmaceutical Co., Tsukuba, Japan) for providing BQ123 and BQ3020. This study was in part supported by Grants-in-Aid from the Ministry of Education, Science, and Culture of Japan (nos. 70574 and 02670383 to S.U., and 02404036 and 03354019 to K.K.) and the Uehara Memorial Foundation.

References

1. Inoue A, Yanagisawa M, Kimura S, Kasuya Y, Miyauchi T, Goto K, Masaki T. Proc Natl Acad Sci USA 1989;86:2863–2867.
2. Arai H, Hori S, Aramori I, Ohkubo H, Nakanishi S. Nature 1990;348:730–732.
3. Sakurai T, Yanagisawa M, Takuwa Y, Miyazaki H, Kimura S, Goto K, Masaki T. Nature 1990;348:732–735.
4. Katoh T, Chang H, Uchida S, Okuda T, Kurokawa K. Am J Physiol 1990;258: F397–F402.
5. Naruse M, Uchida S, Ogata E, Kurokawa K. Am J Physiol 1991;261: F720–F725.
6. Takemoto F, Uchida S, Ogata E, Kurokawa K. Am J Physiol 1993;264: F827–F832.
7. Ihara M, Fukuroda T, Saeki T, Nishikibe M, Kojiri K, Suda H, Yano M. Biochem Biophys Res Commun 1991;178:132–137.
8. Ihara M, Saeki T, Fukuroda T, Kimura S, Ozaki S, Patel AC, YM. Life Sci Pharmacol Lett 1992;51: PL47–52.
9. Burg MB, Grantham JJ, Abramow M, Orloff J. Am J Physiol 1966;210:1293–1298.
10. Shimizu T, Yoshitomi K, Nakamura M, Imai M. J Clin Invest 1988;82:721–730.
11. Grynkiewicz G, Poenie M, Tsien RY. J Biol Chem 1985;260:3440–3450.
12. Kurokawa K, Yoshitomi K, Ikeda M, Uchida S, Naruse M, Imai M. Am Heart J 1992;125: 582–588.
13. Uchida S, Takemoto F, Ogata E, Kurokawa K. Biochem Biophys Res Commun 1992;188:108–113.
14. Zeidel ML, Brady HR, Kone BC, Gullans SR, Brenner BM. Am J Physiol 1989;257: C1101–C1107.
15. Tomita K, Nonoguchi H, Marumo F. J Clin Invest 1990;85:2014–2018.
16. Terada Y, Tomita K, Nonoguchi H, Marumo F. J Clin Invest 1992;90:107–112.
17. Kohan DE, Hughes AK, Perkins SL. J Biol Chem 1992;267:12336–12340.
18. Kohan DE. Am J Physiol 1991;261: F221–F226.
19. Kon V, Badr KF. Kidney Int 1991;40:1–12.

Endothelium-Derived Factors and Vascular Functions. T. Masaki, ed.

Establishment of mice carrying the mutant endothelin-1 gene by gene targeting

Yukiko Kurihara[1], Hiroki Kurihara [1], Hiroshi Suzuki [1,2], Tatsuhiko Kodama [1], Koji Maemura [1], Ryozo Nagai [1], Hideaki Oda [3], Tomoyuki Kuwaki [4], Wei-Hua Cao [4], Nobuo Kamada [2], Kouichi Jishage [2], Yasuyoshi Ouchi [5], Sadahiro Azuma [6], Yutaka Toyoda [6], Takatoshi Ishikawa [3], Mamoru Kumada [4], and Yoshio Yazaki [1,*]

[1]Third Department of Internal Medicine, [3]Department of Pathology, [4]Department of Physiology, [5]Department of Geriatrics, Faculty of Medicine, University of Tokyo, Hongo, Bunkyo-ku, Tokyo 113, [2]Chugai Pharmaceutical Co., Ltd., Komakado, Gotenba, Shizuoka 412, [6]Department of Reproductive and Developmental Biology, Institute of Medical Science, University of Tokyo, Shirokanedai, Minato-ku, Tokyo 108, Japan

Summary

To elucidate the physiological and pathological role of endothelin-1 (ET-1), we designed to establish mice deficient for the ET-1 gene by gene targeting. We disrupted the ET-1 gene in ES cells by homologous recombination. One of the ES clones gave germ-line chimeras which transmit the ET-1 mutation to their offspring. Mice heterozygous for the ET-1 mutation was normal in appearance and fertile, whereas mice homozygous for the ET-1 mutation, in which no ET-1 is produced, was lethal at perinatal stage. Further analysis of their phenotype will reveal the physiological and pathological implication of ET-1.

Introduction

Endothelin-1 (ET-1) is a 21-amino acid vasoconstrictor peptide which is mainly produced by vascular endothelial cells [1]. Three isopeptides (ET-1, 2, and 3) exists and share common receptors with different affinities [2]. ET-1, the only isopeptide which is produced by endothelial cells, have not only the vasoconstrictive and mitogenic effects on vascular smooth muscle cells [1,3] but also various effects such as stimulation of the release of endothelium-derived relaxing factor/nitric oxide (EDRF/NO) and prostacyclin [4], stimulation of the secretion of atrial natriuretic peptide [5] and aldosterone [6], inhibition of renin release [7] and modulation of the central nervous activity [8]. However, the physiological and pathological role of ET-1 remains little known.

---

*To whom correspondence should be addressed.

For clarification of the physiological and pathological role of a specific gene product, knockout mice made by gene targeting serve as one of the best tools because the knockout mice directly reflect deficiency in the specific gene. In this study, we disrupted the ET-1 gene in mouse embryonic stem (ES) cells by homologous recombination to create mice deficient in ET-1 and examined the phenotype of ET-1-deficient mice to elucidate the physiological role of ET-1.

Methods

Cloning of the mouse ET-1 gene and construction of targeting vector

The mouse ET-1 genomic DNA clone containing exon 1 to 5 was isolated from BALB/c mice genomic library in EMBL3 using synthetic oligonucleotide probes corresponding to the mature ET-1 peptide. The targeting vector was constructed from the 7.5-kb EcoRI-EcoRI fragment containing exon 2 to 5 in pBluescriptSK+ (Stratagene). We inserted a 1.8-kb XhoI-SacI fragment of pGK1neo-polyA (gift of M.R. Capecci) into the AccI site in the exon 2 and flanked a 1.8-kb XhoI-SalI fragment of HSV-tk gene (gift of M.R. Capecchi) 3′-downstream to the ET-1 sequence. The resulting plasmid contained 0.9 kb of ET-1 sequence homology at its 5′ end and 6.6 kb at its 3′ end.

Transfection and selection of ES cells

The A3.1 ES cell line derived from the 129/SvJ strain [12] was grown on the feeder layer of mitomycin C-treated G418-resistant mouse embryonic fibroblasts. The cells were electrophoreted with the linealized targeting vector. The cells were then grown on the feeder layer in medium containing G418 and gancyclovir. Then, surviving ES cell colonies were picked up and genomic DNA was subjected to PCR analysis and Southern analysis. Genomic DNA samples which gave a diagnostic band in PCR were analyzed by Southern blot. ES cell clones which represent the ET-1 mutant allele were used for blastocyst injection.

Generation of chimeric mice

Chimeric mice were obtained by injection of ES cells into the C57BL/6J blastocysts at 3.5 dpc. Injected blastocysts were transferred to the uteri of pseudopregnant recipient mice. Chimera were identified by the white and agouti coat color. Male chimeras were mated with ICR mice to obtain F1 offspring with germline transmission. Offspring with white and agouti coat color were genotyped 3 weeks after birth.

Southern blot analysis for genotyping

Genomic DNA was prepared from cultured cells or the tails of newborns. DNA samples were digested with SacI, electrophoreted through 0.7% agarose gels, transferred to nylon membranes and hybridized to the probes at high stringency. The signals were analyzed by autoradiography. The 3.1 kb SacI fragment was diagnostic for the ET-1 mutant allele.

RT-PCR analysis

Total RNA was extracted from whole newborn bodies using RNAzol (BIOTEX) and was reverse-transcribed. PCR was performed on the cDNA samples using a GeneAmp PCR kit (Perkin Elmer Cetus) and primers. Sense (5′-TGTCTTGGGAGC-CGAACTCA-3′) and antisense (5′-GCTCGGTTGTGCGTCAACTTCTGG-3′) primers were deduced from sequence of exon 2 and exon 5 of mouse ET-1 genomic DNA, respectively. Twenty-eight cycles (95°C for 1 min, 60°C for 1 min, 72°C for 1 min) were used to amplify 537 bp products, which were electrophoreted in 1.5% agarose gels and stained with ethidium bromide.

ELISA for the measurement of ET-1 peptide

For the measurement of the ET-1 levels in the lung, we used a sandwich-ELISA as previously described [15]. The lung extracts were prepared from 18.5 dpc newborn littermates [16]. The assay detects as little as 0.2 pg/well of ET-1.

Results

Figure 1 shows the structure of the targeting vector. This was constructed from a 7.5 kilobase (kb) fragment containing the exons 2 to 5 of the mouse ET-1 gene. Exon 2, which encodes the mature 21-amino acid ET-1 peptide [10,11] and our unpubl. data), was disrupted by insertion of the neomycin resistance gene (*neo^r*) from the plasmid pGK1neo. The genomic fragment was flanked at the 3′ end by the herpes simplex virus thymidine kinase gene (*tk*) from the plasmid pMC1tk. A3.1 ES cells [12] were electrophoreted with the linearized targeting vector and survived clones in the positive-negative selection were screened by polymerase chain reaction (PCR) and Southern blot analysis. Figure 2 demonstrates an example of Southern analysis for the ET-1 mutation. Of 281 survived ES clones, 35 clones were proved to be targeted for the ET-1 gene.

One of the targeted clones gave nine male chimeric mice. The degree of chimerism was approximately 10% through 95% ES cell contribution, which is inferred from the ratio of white and agouti versus black coat color. Breeding of the chimeric males to ICR females resulted in germline transmission in two of these chimeras whose progeny inherited the mutant ET-1 allele.

Mice heterozygous for the ET-1 mutation appear normal and were fertile. To examine whether we can obtain viable ET-1$^{-/-}$ homozygotes, we intercrossed ET-1$^{+/-}$

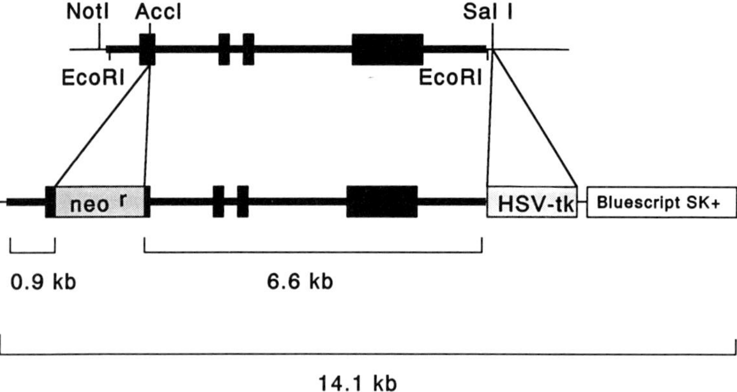

Fig. 1. Construction of the ET-1 targeting vector.

heterozygotes. When 27 offspring from heterozygous crosses were genotyped at 3 weeks of age, 18 were heterozygous and 9 were wild-type. No viable homozygote was found, indicating that the absence of functional ET-1 causes prenatal lethality.

To confirm that no ET-1 is produced in ET-1$^{-/-}$ homozygotes, we examined ET-1

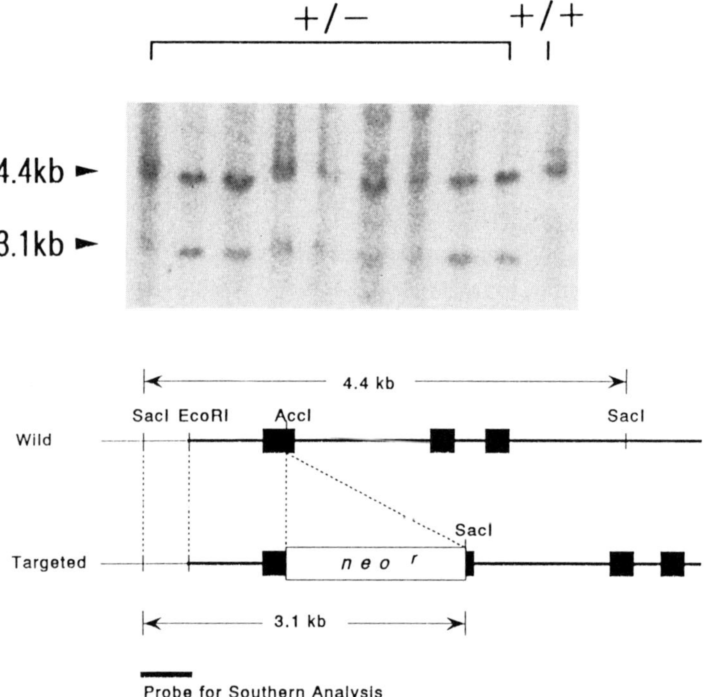

Fig. 2. Southern analysis of the targeted ET-1 gene in ES cells. The 4.4 kb bands indicate the wild-type allele. The 3.1 kb bands indicate the mutated allele. The restriction map of the ET-1 gene and the probe used for Southern analysis are shown in the lower panel.

gene expression by RT-PCR and ET-1 peptide production by ELISA. As expected, no ET-1 mRNA was detected in ET-1$^{-/-}$ homozygotes by RT-PCR (data not shown). ELISA also detected no ET-1 in the lung tissue extracts of ET-1$^{-/-}$ homozygotes, whereas the levels of ET-1 in the lung of ET-1$^{+/-}$ heterozygotes were about 50% of those of wild-type mice (data not shown).

Discussion

In this study, we mutated the ET-1 gene in mouse ES cells by gene targeting and established mice bearing the mutant ET-1 allele. The frequency of 35 targeting in 281 surviving ES colonies in the positive-negative selection was comparative to the highest frequency of the homologous recombination in mouse ES cells (see for example [9,13,14]). The frequency of homologous recombination is greatly dependent on the targeted gene. A gene containing a hotspot of recombination such as the clustering of rearrangements in the immunoglobulin gene is thought to be susceptible to homologous recombination. Another screening of the ET-1 targeting in ES cells gave a similar frequency (data not shown), suggesting that there may be a hotspot of recombination in the mouse ET-1 gene although it is not proved from the sequence.

When the ET-1 targeting vector was constructed, the sequence of exon 2, which encodes mature ET-1 peptide, was disrupted by the neomycin resistance gene so that no protein containing the ET-1 peptide sequence is translated. This is confirmed by the fact that no authentic ET-1 mRNA and peptide were detected in ET-1$^{-/-}$ homozygote tissue by RT-PCR and ELISA.

Although ET-1$^{+/-}$ heterozygotes, in which ET-1 production was decreased by about 50%, seemed normal and were fertile, ET-1$^{-/-}$ homozygotes were lethal before or just after birth. This result suggest that ET-1 is essential for normal fetal development. It may be interesting to investigate the anatomical phenotype of ET-1$^{-/-}$ homozygotes and to clarify the role of ET-1 in the fetal development.

This study has provided mice deficient in vasoactive substance by gene targeting for the first time. The analysis of the phenotype will give us new insights into investigation of not only cardiovascular disease but also ontogeny. For example, the investigation of the involvement of ET-1 in ontogeny may clarify a novel biological activity of ET-1. It is also intriguing to study the hemodynamic changes in ET-1$^{+/-}$ heterozygotes, because whether ET-1 is involved in the pathophysiology of hypertension or vasospasm is still controversial. The measurement of hemodynamic parameters and hormone levels will reveal the role of ET-1 in the maintenance of cardiovascular homeostasis. Furthermore, the ET-1-targeted mutant mice may serve as good disease models when various interventions such as balloon injury of vascular intima are applied. By analyzing such model mice, we can elucidate the involvement of ET-1 in a variety of diseases. Thus, the ET-1 mutant mice will be a good tool for the research including vascular biology and developmental biology.

92

Acknowledgements

We thank M.R. Capecchi for plasmids, K. Miyagawa for mouse genomic library, C. Fujinami, M. Kitsukawa, M. Sasaki, S. Iwasaka, Y. Yamazaki, G.-Y. Ling, and S. Kim for technical assistance, and R.M. Lawn for a critical reading of the manuscript. Y.K. is a Research Fellow of Project Research Subcommittee, Japan Health Sciences Foundation. This work was supported by Grant-in-Aid for Scientific Research from the Ministry of Education, Science and Culture, Japan, the Mitsubishi Foundation, and Uehara Memorial Foundation.

References

1. Yanagisawa M et al. Nature 1988;332:411–415.
2. Inoue A et al. Proc Nat Acad Sci USA 1989;86:2863–2867.
3. Komuro I et al. FEBS Lett 1988;238:249–252.
4. De Nucci GDR et al. Proc Natn Acad Sci USA 1988;85:9797–9800.
5. Fukuda Y et al. Biochem Biophys Res Commun 1988;155:167–172.
6. Morishita R, Higaki J, Ogihara T. Biochem Biophys Res Commun 1989;160:628–632.
7. Rakugi H et al. Biochem Biophys Res Commun 1988;155:1244–1247.
8. Yanagisawa M, Masaki T. Trends in Pharmacol Sci 1989;10:374–378.
9. Mansour SL, Thomas KR, Capecchi MR. Nature 1988;336:348–352.
10. Bloch KD et al. J Biol Chem 1989;264:10851–10857.
11. Inoue A et al. J Biol Chem 1989;264:14954–14959.
12. Azuma S, Toyoda Y. Jpn J Anim Reprod 1991;37:37–43.
13. Zijlstra M et al. Nature 1989;342:435–438.
14. Chisaka O, Musci TS, Capecchi MR. Nature 1992;355:516–520.
15. Suzuki N, Matsumoto H, Kitada C, Masaki T, Fujino M. J Immunol Methods 1989;118:245–250.
16. Matsumoto H, Suzuki N, Onda H, Fujino M. Biochem Biophys Res Commun 1989;164:74–80.

Endothelium-Derived Factors and Vascular Functions. T. Masaki, ed.

Regulation of endothelin production by estrogen: effects of ovariectomy and estrogen replacement in rats

Yasuyoshi Ouchi, Masahiro Akishita, Koichi Kozaki, Seungbum Kim, Michiro Ishikawa, Kenji Toba, and Hajime Orimo
Department of Geriatrics, Faculty of Medicine, University of Tokyo, Tokyo 113, Japan

Summary

This study aims at investigating the role of estrogen in the regulation of endothelin-1 (ET-1) production in rats. Female Wistar rats (8-weeks old) were divided into sham-operated group (Sham), ovariectomized group (OVX), and ovariectomized and estradiol-replaced group. Plasma ET-1 concentration was significantly higher in OVX than that in Sham, and the concentration was comparable to that in the age-matched male rats. Estrogen replacement decreased it to the level in Sham. Metabolic clearance rates for ET-1, calculated by dividing infusion rate of ET-1 by increment in plasma ET-1 concentration, were similar for all groups of rats. These results indicate that estrogen regulates the production of ET-1 in vivo.

Introduction

The lower incidence of atherosclerotic diseases in women before menopause than that in men is an established epidemiological observation [1]. It has been reported that in some models of experimental atherosclerosis females are more resistant to the development of atherosclerosis. For example, Adams et al. [2] demonstrated that, by high cholesterol-feeding, female monkeys develop less coronary atherosclerotic lesions than male monkeys. The similar finding has been reported in some experimental forms of hypertension [3-5]. The mechanism responsible for these gender-related differences has not been clarified; however, the role of estrogen might be important because estrogen replacement therapy protects postmenopausal women from atherosclerotic diseases [6] and, also, inhibits the development of experimental atherosclerosis [2,7] and hypertension [8].

In order to further investigate the mechanism underlying the gender-related difference in the development of atherosclerosis and hypertension, we examined the effect of estrogen on endothelin-1 (ET-1) production in vivo. The reasons why we focused on ET-1 were that ET-1 is not only a potent vasoconstrictor [9] but also a potent mitogen for vascular smooth muscle cells [10], and that plasma concentration of ET-1 is higher in patients with atherosclerotic diseases [11] and essential hypertension [12] than those in the control subjects.

Materials and Methods

Experiment 1: Gender-related difference in plasma ET-1 level

Eight-week old female Wistar rats ($n = 21$) and age-matched male Wistar rats ($n = 5$) were used in this study. The rats were kept in individual stainless steel cages in a room where lighting was controlled (12 hours on, 12 hours off) and room temperature was kept at around 22°C. The animals were fed a commercially available rat chow (MF, Oriental Yeast Co., Tokyo, Japan) and given tap water ad libitum.

The female rats were randomly divided into three groups. The two groups of rats were ovariectomized and the rest of the rats received sham operation. After 1 week of recovery, one group of ovariectomized rats ($n = 7$) received subcutaneous injection of estradiol dipropionate (20 µg/kg; Teikoku Hormone Co. Ltd., Tokyo, Japan) suspended in corn oil once a week (OVX + E group, $n = 7$) for 2 weeks. The other group of ovariectomized rats (OVX group, $n = 7$) and age-matched male rats received the same amount of corn oil as a vehicle once a week for 2 weeks. Blood (3 ml) was collected from each rat in a conscious state through an arterial catheter placed in the left femoral artery into a plastic syringe, and 2 ml of the blood was then transferred into a plastic tube containing EDTA disodium (4 mg) and aprotinin (600 Kallikrein Inactivating Units, KIU). The rest of blood was used for the determination of serum estradiol concentration. Blood samples were centrifuged at $1,000 \times g$ for 10 min at 4°C, and plasma was stored at −80°C until the time of ET-1 and estradiol assays.

Experiment 2: Metabolic clearance rate of ET-1

The metabolic clearance rate (MCR) of ET-1 was determined according to the method described elsewhere [5]. The procedures of sham operation, ovariectomy and estrogen replacement ($n = 5$ for each group) in female Wistar rats were similar to that described in Experiment 1. Age-matched male Wistar rats ($n = 4$) were again used. Two weeks after the start of estradiol or vehicle administration, a catheter (PE-50) was placed in the abdominal aorta through the left femoral artery and another catheter (PE-50) was placed in the right jugular vein under ether anesthesia. The catheters were filled with heparinized saline (100 U/ml) and were sealed and exteriorized at the nape of the neck.

Twenty to 24 hours after surgery, the rats were brought to the laboratory and placed individually in $25 \times 7 \times 7.5$-cm (internal diameter) opaque plastic boxes. These boxes allowed the rats to move backward and forward and to turn around. The arterial catheter was connected to a strain gauge transducer (Statham P10EZ, Gould, Oxnard, CA, USA) to measure mean arterial blood pressure (MABP) and heart rate (HR). A slot along the top of the box allowed free movement of the catheter. MABP and HR were recorded on a thermal pen recorder (WR3701, Graphtec, Tokyo, Japan) throughout the experiment.

After the rats had become accustomed to the surroundings for 30 to 45 min and

MABP and HR had become stabilized, an initial blood sample (2 ml) was slowly drawn from the arterial catheter into a plastic syringe, and then transferred into a plastic tube containing EDTA disodium (4 mg) and aprotinin (600 KIU). An equal volume of rat donor blood, prewarmed to 37°C, was simultaneously injected through a venous catheter. Donor blood was obtained immediately before the experiment from male and female rats by decapitation. Each rat received donor blood from rats of the corresponding sex. After the initial blood sampling, ET-1 (Sigma, St. Louis, MO, USA) dissolved in phosphate-buffered saline containing 0.1% bovine serum albumin was infused intravenously at a rate of 200 pg/kg/min for 30 min by using Microinjection Pump (CMA 100, Carnegie Medicin AB, Stockholm, Sweden). The rate of ET-1 infusion was then increased to 600 pg/kg/min and the infusion was continued for another 30 min. At the end of each 30-min period, additional blood samples (2 ml each) were drawn with the simultaneous replacement of donor blood. Blood samples were centrifuged and plasma was stored as described in Experiment 1 until the time of ET-1 assay.

ET-1 and estradiol assay

Plasma (1 ml) was diluted four times by adding 3 ml of 4% acetic acid solution. The diluted plasma was extracted with Sep-Pak C18 cartridges (Water Associates, Milford, MA, USA) preconditioned sequentially with 86% ethanol containing 4% acetic acid, methanol, distilled water and finally 4% acetic acid. ET-1 was extracted by adding 1 ml of 4% acetic acid. The extracts were evaporated by blowing nitrogen gas for 5 to 6 hours at 37°C, and were then reconstructed in 1 ml ethanol. The solution was evaporated again by blowing nitrogen gas for approximately 30 min at 37°C. The residue was dissolved in 250 μl of 50 mM phosphate buffer solution (pH 7.0), sonicated for 10 min, and centrifuged at $9,200 \times g$ for 10 min. The supernatant was used for ET-1 assay.

ET-1 concentration in the supernatant was measured in duplicate by Sandwich-enzyme immunoassay according to the method described by Suzuki et al. [13]. The monoclonal antibody raised against synthetic human ET-1 used in this study has been reported to show cross-reactivities of 100% for ET-1, 160% for ET-2, 0.21% for ET-3 and 0.16% for big ET-1. The detection sensitivity was 0.5 pg/ml for plasma samples. Serum estradiol level was measured with specific radioimmuno-assay using double antibody technique.

Calculation of MCR for ET-1

The MCR of ET-1 was calculated by dividing each infusion rate of ET-1 by the increment of plasma ET-1 level; that is, plasma ET-1 level at the end of each infusion period minus plasma ET-1 level before the start of ET-1 infusion.

96

Statistics

Data were analyzed using one-factor analysis of variance. If statistically significant effects were found, Newman-Keuls test was performed to isolate the differences between groups. A p value of <0.05 was considered to be significant. All data in the text, table and figures are presented as means ± SEM.

Results

Experiment 1: Gender-related difference in plasma ET-1 level

As shown in Table 1, plasma ET-1 level was significantly lower in sham-operated female rats than that in male rats. Ovariectomy significantly increased ET-1 level to that observed in male rats, and estrogen replacement decreased it to the level in sham-operated female rats. Estrogen replacement significantly increased serum estradiol level in ovariectomized rats.

Table 1.   Plasma endothelin-1 (ET-1) and estradiol (E_2) levels in sham-operated (Sham), ovariectomized (OVX), estradiol-replaced ovariectomized (OVX + E), and age-matched male (Male) rats.

	n	ET-1 (pg/ml)	E_2 (pg/ml)
Sham	7	0.68 ± 0.14[b,c]	34.6 ± 8.5
OVX	7	1.32 ± 0.14	11.2 ± 1.6
OVX + E	7	0.85 ± 0.12[a]	21.7 ± 4.2[a]
Male	5	1.49 ± 0.31	n.d.

n = number of rats, n.d. = not determined. [a]p<0.05, [b]<0.01 vs. OVX, [c]p<0.05 vs. Male.

Experiment 2: Metabolic clearance rate of ET-1

As shown in Fig. 1 and Fig. 2, ET-1 infusion did not affect MABP and HR. As shown in Fig. 3, plasma ET-1 concentrations before ET-1 infusion showed the similar gender-related difference as observed in Experiment 1. The effects of ovariectomy and estrogen replacement were also similar. Plasma ET-1 concentration in each group of rats increased in a dose-dependent manner by ET-1 infusion. As shown in Fig. 4, no significant differences in calculated MCR were observed among four groups of rats both at 200 and 600 pg/kg/min of ET-1 infusion.

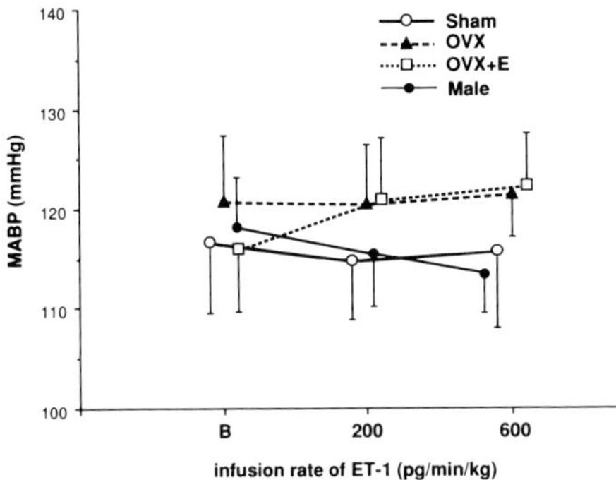

Fig. 1. Mean arterial blood pressure (MABP) before (B) and after infusion of graded doses of endothelin-1 (ET-1). Sham = sham-operated female rats (n = 5), OVX = ovariectomized female rats (n = 5), OVX + E = estradiol-replaced ovariectomized rats (n = 5), Male = age-matched male rats (n = 4).

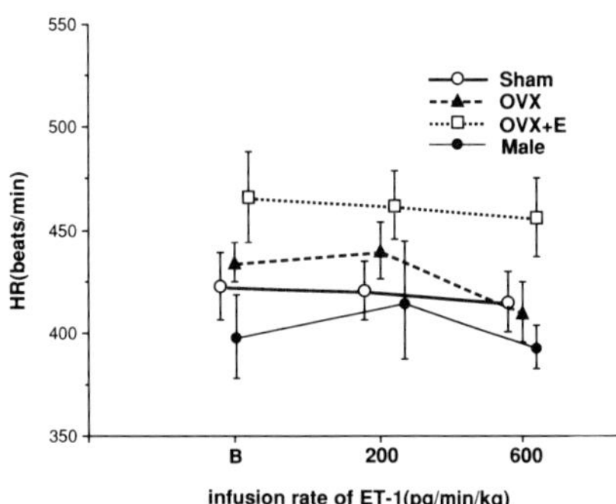

Fig. 2. Heart rate (HR) before (B) and after infusion of graded doses of endothelin-1 (ET-1). Sham = sham-operated female rats (n = 5), OVX = ovariectomized female rats (n = 5), OVX + E = estradiol-replaced ovariectomized rats (n = 5), Male = age-matched male rats (n = 4).

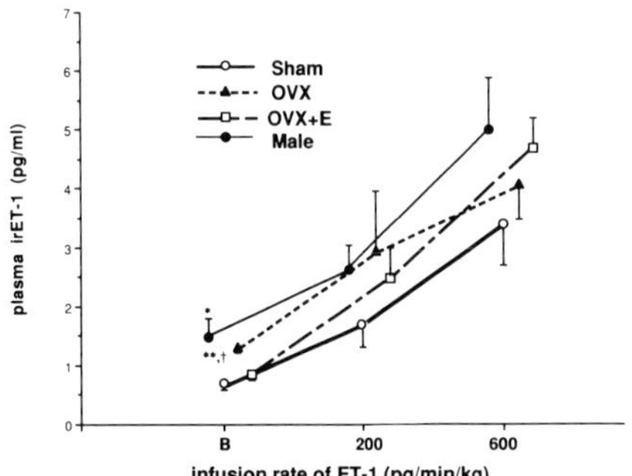

Fig. 3. Plasma immunoreactive endothelin-1 (irET-1) levels before (B) and after infusion of graded doses of endothelin-1 (ET-1). Sham = sham-operated female rats (n = 5), OVX = ovariectomized female rats (n = 5), OVX + E = estradiol-replaced ovariectomized rats (n = 5), Male = age-matched male rats (n = 4). *p < 0.05, **p < 0.01 vs. OVX, †p < 0.05 vs. male.

Discussion

In the present study, we have shown that plasma ET-1 level is lower in female rats than that in male rats. This gender-related difference in plasma ET-1 level might be associated with the higher plasma level of estrogen in female rats because ovariectomy decreased plasma ET-1 level to that in male rats. Moreover, the fact that estrogen replacement decreased the level to that in female rats further supports the possible role of estrogen in the regulation of plasma ET-1 level. Our findings are compatible with the recent investigations by Miyauchi et al. [14] and Polderman et al. [15]. Both groups of investigators reported that plasma endothelin levels are

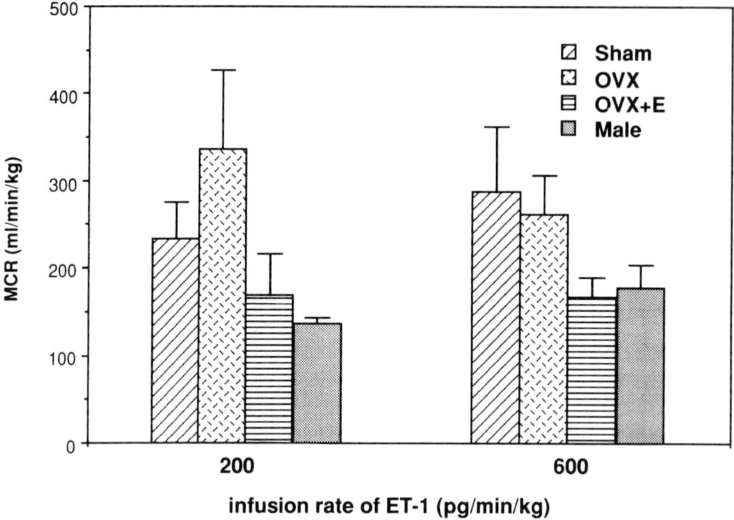

Fig. 4. Metabolic clearance rates (MCR) in sham-operated (Sham; $n = 5$), ovariectomized (OVX; $n = 5$), estradiol-replaced ovariectomized (OVX + E; $n = 5$), and age-matched male (Male; $n = 4$) rats.

higher in men than those in women. Moreover, Polderman et al. [15] suggested the role of gonadal hormones on the regulation of plasma endothelin levels by showing that the levels decrease in male-to-female transsexuals and increase in the opposite transsexuals.

The gender-related difference in plasma ET-1 level could be a consequence either from the difference in production or degradation of ET-1 in vivo. To investigate this issue, we determined the effects of ovariectomy and estrogen replacement on MCR of ET-1, and we found that MCR of ET-1 was not influenced by gender, ovariectomy and estrogen replacement. This finding indicates that the gender-related difference in plasma ET-1 level is attributed to the difference in the production of ET-1. To our knowledge, this is the first demonstration that estrogen modulates the production of ET-1 in vivo. The sources of ET-1 in systemic circulation have been considered to be vascular endothelial cells, vascular smooth muscle cells and the pituitary gland. We were unable to determine the site for ET-1 release responsible for the gender-related difference in plasma ET-1 level. This issue should be investigated by using cultured vascular endothelial cells, cultured smooth muscle cells and organ culture system of the hypothalamus/pituitary gland.

The mechanism underlying the effect of estrogen on ET-1 production remains to be determined. This might be mediated by estrogen receptors because functional estrogen receptors have been identified in cultured vascular smooth muscle cells [16] and vascular endothelial cells [17]. Once estrogen binds to estrogen receptors in the target cells, estrogen-receptor complex translocates in the nucleus, binds to estrogen responsive element on genomic DNA, and some target genes are expressed [18]. Since estrogen acts to inhibit ET-1 production as demonstrated in the present study, it is possible that some inhibitory factors of ET-1 production intervene between target gene of estrogen and endothelin gene. Another possibility is an inhibition of conversion from big endothelin to mature endothelin molecule through the inhibitory effect of estrogen on endothelin-converting enzyme activity. Further investigation is needed to clarify this point.

Evidence has accumulated indicating that endothelin is involved in the development of atherosclerosis because the presence of endothelin in the atherosclerotic lesion has been immunohistochemically demonstrated [11]. Moreover, it has been reported that there is a significant correlation between plasma ET-1 level and the extent of advanced atherosclerosis [11]. Proliferation of vascular smooth muscle cells in the intima of blood vessels is a major event in the course of atherogenesis [19] and endothelin has been reported to actually be a potent mitogen for vascular smooth muscle cells [10]. The gender-related difference in ET-1 production demonstrated in this study might be one possible mechanism of gender-related difference in atherogenesis. In conclusion, estrogen has an inhibitory effect on the production of ET-1 in vivo, and this might be involved in the mechanism underlying the protection of premenopausal women from atherosclerosis.

Acknowledgement

This work was supported by grants from the Uehara Memorial Foundation and the Sankyo Life Science Research Foundation.

References

1. Lerner DJ, Kannel WB. Am Heart J 1986;111:383–390.
2. Adams MR, Clarkson TB, Kaplan JR, Koritnik DR. Transplantation Proc 1989;21: 3662–3664.
3. Iams SG, Wexler BC. J Lab Clin Med 1977;90:997–1003.
4. Dahl LK, Knudsen KD, Ohanian EV, Muirhead M, Tuthill R. J Exp Med 1975;142:748–759.
5. Ouchi Y, Share L, Crofton JT, Iitake K, Brooks DP. Hypertension 1987;9:172–177.
6. Stampfer MJ, Willet WC, Colditz GA, Rosener B, Speizer FE, Henekens CH. New Engl J Med 1985;313:1044–1049.
7. Clarkson TB, Adams MR, Kaplan JR, Shively CA, Koritnik DR. Am J Obstet Gynecol 1989;160:1280–1285.
8. Iams SG, Wexler BC. J Lab Clin Med 1979;94:608–616.
9. Yanagisawa M, Kurihara H, Kimura S, Tomobe M, Kobayashi Y, Mitsui Y, Yazaki Y, Goto K, Masaki T. Nature 1988;332:411–415.
10. Komuro I, Kurihara H, Sugiyama T, Yoshizumi M, Takaku F, Yazaki Y. FEBS Lett 1988;238:249–252.
11. Lerman A, Edwards BS, Hallett JW, Heublein DM, Sandberg SM, Burnett JC Jr. New Engl J Med 1991;325:997–1001.
12. Saito Y, Nakao K, Mukoyama M, Imura H. New Engl J Med 1990;322:205.
13. Suzuki N, Matsumoto H, Kitada C, Masaki T, Fujino M. J Immunol Methods 1989;118:245–250.
14. Miyauchi T, Yanagisawa M, Iida K, Ajisaka R, Suzuki N, Fujino M, Goto K, Masaki T, Sugishita Y. Am Heart J 1992;123:1092–1093.
15. Polderman KH, Stehouwer CDA, van Kamp GJ, Dekker GA, Verheugt FWA, Gooren LJG. Ann Intern Med 1993;118:429–432.

16. Orimo A, Inoue S, Ikegami A, Hosoi T, Akishita M, Ouchi Y, Muramatsu M, Orimo H. Biochem Biophys Res Commun 1993;195:730–736.
17. Colburn P, Buonassisi V. Science 1978;201:817–819.
18. Clark JH, Schrader WT, O'Malley BW. In: Wilson JD, Foster DW (eds.) Williams Textbook of Endocrinology 8th Edition. Philadelphia: Sounders, 1992:35–90.
19. Ross R. Nature 1993;362:801–809.

Vascular remodeling factors

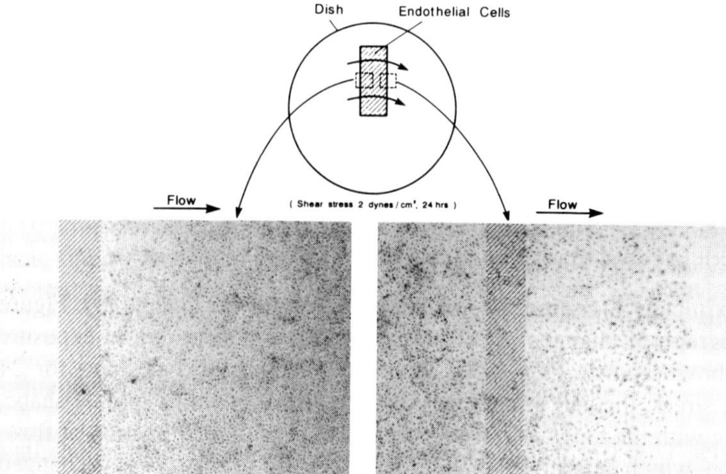

Fig. 3. The effect of shear stress on the migration and proliferation of ECs.

on the effect of shear stress on proliferation of cells not in the process of regeneration to such an EC-free site but in the confluent situation have been reported. No definite conclusion has yet been drawn concerning these observations.

ECs produce and secrete various physiological activators, and their functions are affected by shear stress. Production of histamine (increasing endothelial permeability) [10], prostaglandin I_2 (PGI_2) and endothelium-derived relaxing factor (EDRF), which have vasodilative and platelet aggregation-inhibiting action, and tissue plasminogen activator (tPA) involved in thrombolysis is stimulated by shear stress. It has been revealed by determination of 6-keto prostaglandin $F_{1\alpha}$ in conditioned medium that PGI_2 is immediately released in a burst when shear stress is applied to cultured EC in the flow-loading device and the amount of PGI_2 produced is decreased with time [11]. This effect is more marked in the pulsatile flow than in the steady flow.

EDRF is believed to be nitric oxide (NO). In our recent study, to investigate the effect of shear stress on NO production by ECs, cultured ECs were exposed to shear stress in a parallel plate type of device and examined for changes in NO production. Since NO activates guanylate cyclase and increases cGMP levels in ECs, cGMP levels in ECs were used as an indication of the amount of NO produced. Application of flow increased the cGMP levels in proportion to the magnitude of shear stress (Fig. 4). From the fact that the shear-induced increase in cGMP is completely inhibited by treatment of ECs with a specific inhibitor of NO production, N^w-methyl L-arginine, the substance increasing cGMP is considered to be NO. That is, NO is produced from ECs shear stress-dependently.

ECs express various adhesion molecules on the cell surface and adhere to circulating leukocytes. The adhesive interaction is an important event in the development of tissue injury or vascular disease such as atherosclerosis. A number

Nitrite assay

The nitrite level in cell-free supernatant was determined by the addition of an equal volume of Griess reagent (1% sulfanilamide and 0.1% naphthylethylenediamine dihydrochloride in 2% phosphoric acid). The absorbance at 540 nm was measured, and nitrite concentration was determined using sodium nitrite as a standard.

Determination of DNA synthesis

Confluent VSMC or co-cultured cells were incubated with IL-1 or various drugs for the time indicated and labelled with 2 μCi/well of [^{3}H]thymidine during the last 4 h period of the incubation. For VSMC, which were cultured in 12-well plates, the cell layers were then washed three times with cold phosphate buffered saline (PBS) and treated with 5% trichloroacetic acid and ethanol-ethylether (3:1, v/v). The residues in the wells were solubilized in 0.3N NaOH, and the radioactivity of aliquots of the solution was measured after the neutralization of pH. For EC, which were cultured on the collagen-coated membrane of the transwells, the cell layers were washed three times with cold PBS and then cells were solubilized in 0.3 N NaOH, and the radioactivity of aliquots of the solution was measured after the neutralization of pH.

Statistical analysis

Statistical analysis was performed by one-way analysis of variance. Results are expressed as mean ± SEM. A value of $p < 0.05$ was considered significant.

Results

Figure 1 shows the kinetics of NO release from VSMC following exposure to 10 ng/ml IL-1 in the presence or absence of 3 mM L-NMMA, an inhibitor of NO synthesis, at a higher concentration than that of L-arginine in the medium (0.4 mM). NO release from VSMC was determined by measuring nitrite levels in the culture supernatant. IL-1 induced a time-dependent release of high levels of NO from VSMC. Coincubation with L-NMMA significantly inhibited IL-1-induced NO release from VSMC.

We next examined whether high levels of NO induced by IL-1 mediated cytotoxicity in VSMC. Incubation with IL-1 for 72 h significantly inhibited [^{3}H]thymidine incorporation in VSMC. However, in the presence of L-NMMA, IL-1 did not inhibit but rather stimulated [^{3}H]thymidine incorporation in VSMC. Coincubation with 30 mM L-arginine inhibited the stimulation of [^{3}H]thymidine incorporation induced by IL-1 in the presense of L-NMMA (Fig. 2). Furthermore, incubation of VSMC with IL-1 for 96 h induced cell detachment and cell death (more than 90% of VSMC were stained with 0.1% trypan blue). However, coincubation with 3 mM L-NMMA prevented this cell damage induced by IL-1 (Fig. 3).

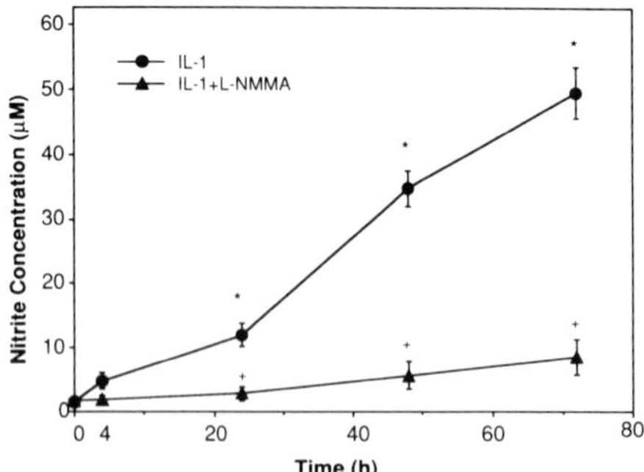

Fig. 1. Kinetics of NO release after stimulation of VSMC with IL-1. Confluent VSMC were preincubated for 48 h with DMEM containing 0.1% BSA, then cells were incubated in new medium containing IL-1 (10 ng/ml) for the times indicated. Nitrite concentration in the medium was determined as described in the text. Values are mean SEM ($n = 5$). *p<0.05, significantly different from control. +p<0.05, significantly different from cells treated with IL-1 alone.

As shown in Fig. 4, when VSMC were co-cultured with EC, incubation with IL-1 for 48 h induced a release of high levels of NO from VSMC co-cultured with EC, as high as that from VSMC alone. IL-1 also induced inhibition of [^{3}H]thymidine incorporation in VSMC co-cultured with EC. Furthermore, coincubation with 10^{-8} M ET-1 significantly enhanced the IL-1-induced inhibition of [^{3}H]thymidine incorporation in VSMC. However, ET-1 did not affect the amount of NO release from co-cultured cells.

Next, we examined whether high levels of NO induced by IL-1 mediated cytotoxity in EC. Incubation with IL-1 for 24 h significantly inhibited [^{3}H]thymidine incorporation in EC alone (Table 1). L-NMMA did not affect this inhibition induced by IL-1. On the other hand, when EC was co-cultured with VSMC, incubation with

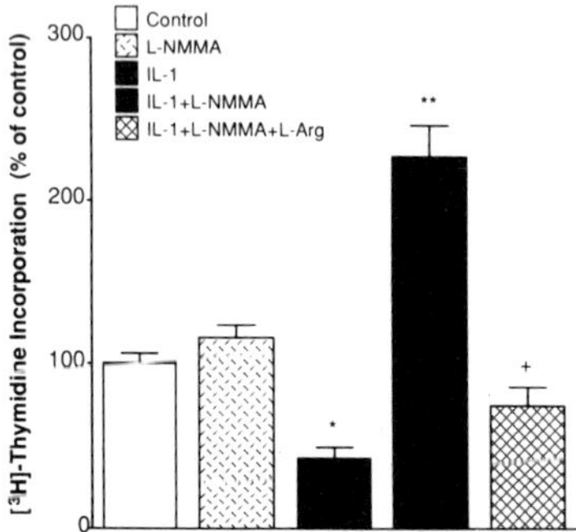

Fig. 2. Effects of IL-1 on [^{3}H]thymidine incorporation of VSMC. Confluent VSMC were incubated for 68 h with IL-1 in the presence or absence of L-NMMA or L-arginine (L-Arg). After collection of the medium for the determination of nitrite concentration, cells were labelled with [^{3}H]thymidine for a further 4 h. [^{3}H]Thymidine incorporation in VSMC was measured as described in the text. Values are mean SEM ($n = 5$). *p<0.05, significantly different from control. **p<0.05, significantly different from cells treated with IL-1 alone. +p<0.05, significantly different from cells treated with IL-1 and L-NMMA.

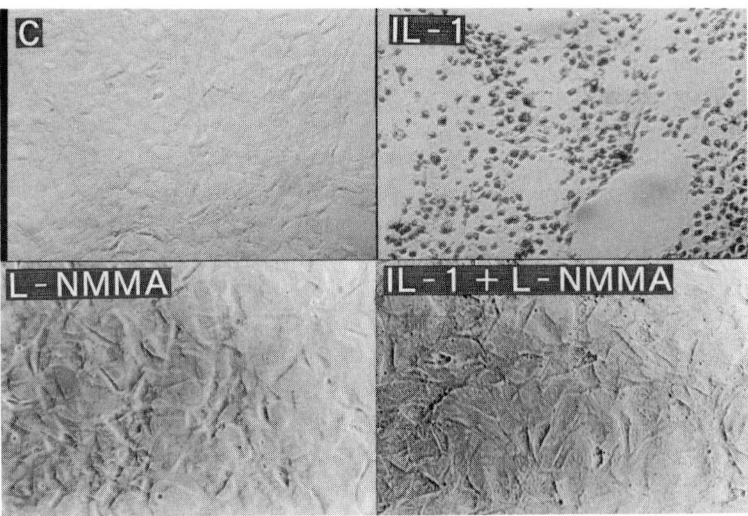

Fig. 3. Cytotoxic effects of IL-1 on VSMC. Confluent VSMC were incubated for 96 h with DMEM containing vehicle or IL-1 (10 ng/ml) in the presence or absence of L-NMMA (3 mM). Then cells were stained with 0.1% trypan blue.

IL-1 for 72 h induced a significant stimulation of [^{3}H]thymidine incorporation in EC (Table 1). Whereas, incubation with IL-1 induced significant inhibition of [^{3}H]thymidine incorporation in VSMC (Fig. 4). However, coincubation with L-NMMA abolished the stimulation of [^{3}H]thymidine incorporation in EC induced by IL-1. Furthermore, the IL-1-induced stimulation of [^{3}H]thymidine incorporation in EC co-cultured with VSMC was partially but significantly inhibited by antibody to bFGF at a concentration of 10 µg/ml, whereas non-immune IgG had no effect (Fig. 5).

Table 1. Effects of IL-1 on [^{3}H]thymidine incorporation in EC co-cultured with VSMC.

	EC (% of control)	EC + VSMC (% of control)
Control	100.0±6.9	100.0± 8.2
L-NMMA (3 mM)	111.7±8.2	121.5±11.8
IL-1 (10 ng/ml)	48.6±7.1*	293.3±14.4**
IL-1 + L-NMMA	50.1±6.6	57.7± 7.9***

Confluent EC were co-cultured in DMEM containing 10 ng/ml IL-1 in the presence or absence of L-NMMA with or without L-arginine (L-Arg) for 68 h. Cells were labelled for a further 4 h with [^{3}H]thymidine. [^{3}H]thymidine incorporation in EC were determined as described in the text. Values are mean SEM (*n* = 6). *p<0.05, significantly different from control in EC alone. **p<0.05, significantly different from EC alone treated with IL-1. ***p<0.05, significantly different from EC co-cultured with VSMC treated with IL-1 alone.

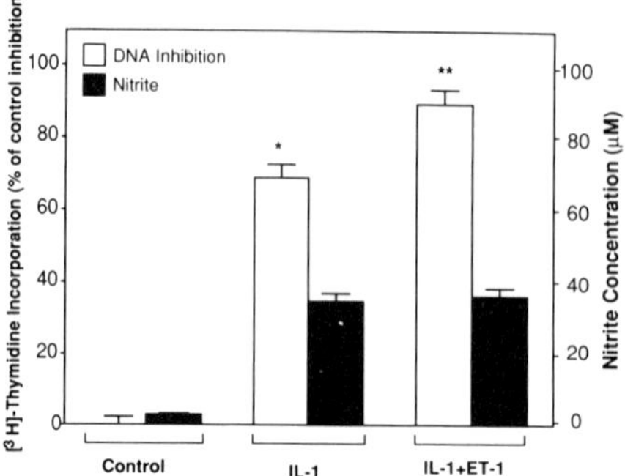

Fig. 4. Effects of IL-1 and ET-1 on NO release and [^{3}H]thymidine incorporation of VSMC co-cultured with EC. Confluent VSMC and EC were co-cultured in DMEM containing 10 ng/ml IL-1 or 10^{-8} M ET-1 for 68 h. After collection of the medium for the determination of nitrite concentration, cells were pulse labelled with [^{3}H]thymidine for a further 4 h. [^{3}H]Thymidine incorporation in VSMC was measured as described in the text. Values are mean SEM ($n = 5$). *p<0.05, significantly different from control. **p<0.05, significantly different from cells treated with IL-1 alone.

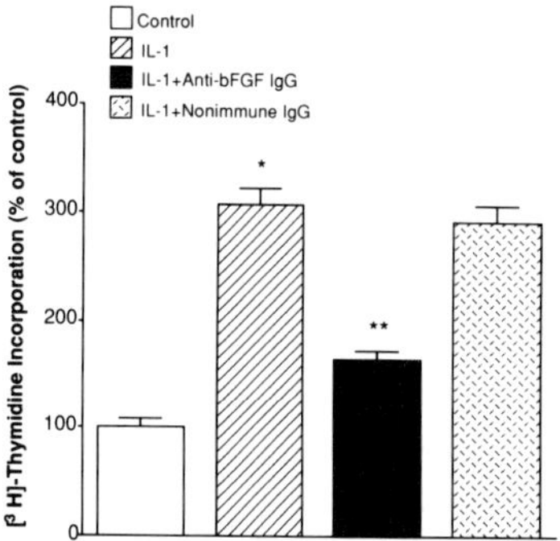

Fig. 5. Effects of anti-bFGF IgG on the stimulation of [^{3}H]thymidine incorporation induced by IL-1 in EC co-cultured with VSMC. Confluent EC were co-cultured with VSMC in DMEM containing IL-1 in the presence or absence of anti-bFGF IgG or non-immune IgG for 68 h. Co-cultured cells were pulse labeled with [^{3}H]thymidine for a further 4 h. [^{3}H]Thymidine incorporation in EC was determined as described in the text. Values are mean ± SEM ($n = 5$). *p<0.05, significantly different from control. **p<0.05, significantly different from cells treated with IL-1 alone.

Discussion

The inducible NO synthase (iNOS) is known to release large amounts of NO for longer periods of time when macrophages are activated by endotoxin or cytokines. High levels of NO released from macrophages may not only be toxic to microbes and malignant cells but may also damage healthy tissue, since high levels of NO impairs the function of mitochondrial and other FeS-containing enzymes which are important in the regulation of cell function and viability [16-18]. Recent evidence is accumulating that NO may participate in inflammatory and autoimmune-mediated tissue destruction in the central nervous system [19] and pancreatic islet cells [20]. In this study we demonstrated that prolonged incubation with IL-1 induced a high level of NO release in the medium and cell damage in VSMC. Furthermore, L-NMMA, an inhibitor of NO synthesis, prevented the cell damage induced by IL-1, suggesting that a high level of NO is involved in the cell damage induced by IL-1 in VSMC.

On the other hand, IL-1 did not inhibit but rather stimulated [^{3}H]thymidine incorporation in VSMC when NO synthesis was inhibited by L-NMMA. Therefore, IL-1 may have both inhibitory and stimulatory signals for DNA synthesis in VSMC. IL-1-induced stimulation of DNA synthesis may be mediated by induction of the platelet-derived growth factor A-chain gene [21].

NO is known to bind with very high affinity to iron in the heme of soluble guanylate cyclase to stimulate the formation of cyclic GMP in VSMC. We have recently shown that NO directly induces the activation of a cyclo-oxygenase, a heme-containing enzyme, and partially mediates IL-1-induced prostaglandin E_2 production [22]. Although prostaglandin E_2 is known to inhibit VSMC growth, co-incubation with indomethacin, a cyclo-oxygenase inhibitor, did not prevent the cytotoxic effects induced by prolonged incubation with IL-1 in VSMC (data not shown), suggesting that prostaglandins may not participate in the cell damage induced by IL-1.

Recent studies have demonstrated that plasma ET-1 concentrations are elevated in patients with symptomatic atherosclerosis [23], and ET-1 mRNA expression is elevated in atherosclerotic lesions [24]. Exogenously introduced ET has been reported to cause not only contraction of various smooth muscle but also gastric mucosal damage [25] and acute renal failure [26]. Furthermore, Watanabe et at. [27] have recently demonstrated that a monoclonal antibody against ET-1 reduced the infarct size after coronary occlusion in rats. In the present study, ET-1 enhanced the IL-1-induced cell damage in VSMC co-cultured with EC. Although the mechanism of the enhancement induced by ET-1 is not clear, ET-1 did not affect the NO release induced by IL-1 in VSMC co-cultured with EC. In this regard, it has been reported that NO amplifies the gene transcription induced by calcium (Ca^{2+}) in neuronal cells [28]. Since ET-1 induces an increase in intracellular Ca^{2+} in VSMC (data not shown), Ca^{2+} might be involved in ET-1-induced enhancement of cell damage induced by IL-1 in VSMC.

As shown in Table 1, IL-1 inhibited [^{3}H]thymidine incorporation in EC alone. These results are consistent with the previous results [29, 30]. In contrast, IL-1 induced the stimulation of [^{3}H]thymidine incorporation in EC, when EC were co-cultured with VSMC. Furthermore, the IL-1-induced stimulation of [^{3}H]thymidine incorporation in EC was inhibited by co-incubation with anti-bFGF IgG, suggesting that bFGF may participate in the mechanism of IL-1 induced stimulation in EC co-cultured with VSMC.

bFGF is known to be a potent EC and VSMC mitogen [31,32] and is synthesized by three cell types associated with atherosclerosis: EC [33], macrophages [34], and VSMC [35]. However, biosynthetic studies indicate that bFGF is a cellular rather than a secreted protein and may be released from cells after injury [36]. In this study, prolonged incubation with IL-1 induced a high level of NO release in the medium and cell damage in both VSMC alone and VSMC co-cultured with EC. Furthermore, L-NMMA which inhibited both NO release and cell damage induced by IL-1, completely inhibited the stimulation of [^{3}H]thymidine incorporation in EC co-cultured with VSMC. These results suggest that bFGF released from damaged VSMC stimulates neighboring EC proliferation in co-culture system. Thus, VSMC damage induced by high levels of NO may be an important trigger of vascular remodeling in atherosclerotic lesions and that NO may be an important signal in the mechanism of the response to injury model [13].

Acknowledgements

We thank Taeko Kaimoto for her excellent technical assistance. This work was supported by a Grant from Uehara Memorial Foundation.

References

1. Ross R. N Engl J Med 1986;314:488–500.
2. Zierler RE, Bandyk DF, Thiele BL, Strandness DE Jr. Arch Surg 1982;117: 1408–1415.
3. McBride W, Lange R, Hills LD. N Engl J Med 1988;318:1734–1737.
4. Bocan TMA, Schifani TA, Guyton JR. Am J Pathol 1986;123:413–424.
5. Ross R. Nature 1993;362:801–809.
6. Moncada S, Palmer RMJ, Higgs EA. Pharmacol Rev 1990;43:109–142.
7. Snyder SH. Science 1992;257:494–496.
8. Dinerman JL, Lowenstein CJ, Snyder SH. Circ Res 1993;73:217–222.
9. Beasley D, Schwartz JH, Brenner BM. J Clin Invest 1991;87:602–608.
10. Bernhardt J, Tschudi MR, Dohi Y, Gut I, Urwyler B, Buhler FR, Luscher TF. Biochem Biophys Res Commun 1991;180:907–912.
11. Nunokawa Y, Ishida N, Tanaka S. Biochem Biophys Res Commun 1993;191:89–94.
12. Stamler JS, Singel DJ, Loscalzo J. Science 1992;258:1898–1902.
13. Ross R. N Engl J Med 1986;314: 488–500.
14. Faggiotto A, Ross R, Harker L. Arteriosclerosis 1984;4:341–356.
15. Fukuo K, Morimoto S, Koh E, Yukawa H, Tsuchiya S, Imanaka S, Yamamoto H, Onishi T, Kumahara Y. Biochem Biophys Res Commun 1986;136:247–253.
16. Drapier J-C, Hibbs JB Jr. J Immunol 1988;140:2829–2838.
17. Albina JE, Caldwell MD, Henry WL Jr, Mills CD. J Exp Med 1989;169:1021–1029.
18. Kolb H, Kolb-Bachofen V. Immunol Today 1992;13:157–159.
19. Koprowski H, Zheng YM, Heber-Katz E, Fraser N, Rorke L, Fu ZF, Hanlon C, Dietzschold B. Proc Natl Acad Sci USA 1993;90:3024–3027.
20. Corbett JA, Wang JL, Sweetland MA, Lancaster JR Jr, McDaniel ML. J Clin Invest 1992;90:2384–2391.
21. Raines EW, Dower SK, Ross R. Science 1989;243:393-396.
22. Inoue T, Fukuo K, Morimoto S, Koh E, Ogihara T. Biochem Biophys Res Commun 1993;194:420-424.
23. Lerman A, Edwards BS, Hallet JW, Heublein DM, Sanberg SM, Burnett JC Jr. N Engl J Med 1991;325:997–1001.
24. Winkles JA, Alberts GF, Brogi E, Libby P. Biochem Biophys Res Commun 1993;191:1081–1088.
25. Whittle BJR, Esplugues JV. Br J Pharmacol 1988;95:1011–1013.

26. Shibata Y, Suzuki N, Shino A, Matsumoto H, Terashita Z, Kondo K, Nishikawa K. Life Sci 1990;46:1611–1618.
27. Watanabe T, Suzuki N, Shimamoto N, Fujino M, Imada A. Circ Res 1991;69:370–377.
28. Peunova N, Enikolopov G. Nature 1993;364:450–453.
29. Norioka K, Hara M, Kitani T, Hirose T, Hirose W, Harigai M, Suzuki K, Kawakami M, Tabata H, Kawagoe M, Nakamura H. Biochem Biophys Res Commun 1987;145:969–975.
30. Cozzolino F, Torcia M, Aldinucci D, Ziche M, Almerigogna F, Bani D, Stern DM. Proc Natl Acad Sci USA 1990;87:6487–6491.
31. Folkman J, Klagsbrun M. Science 1987;235:442–447.
32. Gospodarowicz D, Cheng D, Lui G-M, Baird A, Esch F, Bohlen P. Endocrinology 1985;117:2283-2391.
33. Schweigerer L, Neufeld G, Friedman J, Abraham JA, Fiddes JC, Gospodarowicz D. Nature 1987;325:257–259.
34. Baird A, Mormede P, Bohlen P. Biochem Biphys Res Commun 1985;126: 358–364.
35. Winkles JA, Friesel R, Burgess WH, Howk R, Mehlman T, Weinstein R, Maciag T. Proc Natl Acad Sci USA 1987;84:7124–7128.
36. Gajdusek CM, Carbon S. J Cell Physiol 1989;139:570–579.

Endothelium-Derived Factors and Vascular Functions. T. Masaki, ed.

C-type natriuretic peptide (CNP) as a novel endothelium-derived relaxing peptide: occurrence of vascular natriuretic peptide system

Kazuwa Nakao, Hiroshi Itoh, Takaaki Yoshimasa and Hiroo Imura

Second Division, Department of Medicine, Kyoto University School of Medicine, 54 Shogoin Kawahara-cho, Sakyo-ku, Kyoto 606, Japan

Summary

C-type natriuretic peptide (CNP), the third member of the natriuretic peptide family, has been considered to act mainly as a neuropeptide. We discovered that CNP is produced and secreted by the endothelial cells. The endothelial secretion of CNP is regulated by various vasoactive substances, including growth factors and cytokines. In in vivo blood vessel walls, the gene transcripts of CNP and ANP-B receptor, the specific receptor for CNP were detected in a wide range of mammals. The gene expression of ANP-B receptor was up-regulated in the proliferative vascular lesion. Furthermore, CNP is detected in both human and rat plasma. Exogenously-administered CNP elicited hypotensive and diuretic/natriuretic actions in humans. These results support the existence and significance of "vascular natriuretic peptide system", in which CNP can act as a novel endothelium-derived relaxing peptide affecting vascular tone and growth.

Introduction

The natriuretic peptide system (NPS) comprises three endogenous ligands, atrial natriuretic peptide (ANP), brain natriuretic peptide (BNP), and C-type natriuretic peptide (CNP) and three natriuretic peptide receptors (ANPR-A, ANPR-B, and clearance receptor). Natriuretic peptides occur in the heart and in the central nervous system, where they act as cardiac hormones and neuropeptides, respectively, being responsible for body fluid homeostasis and blood pressure control [1,2]. While ANP and BNP are demonstrated to be secreted from the atrium and the ventricle, respectively [3], we and others elucidated that CNP is mainly produced in the central nervous system but is not detected in the heart [4–6]. Taken together with our finding that the concentrations of CNP-like immunoreactivity in the human brain and human cerebrospinal fluid are much higher than ANP and BNP [7,8] and that CNP elicits the central action [9], CNP is considered to be the major natriuretic peptide in the central nervous system and to act as a neuropeptide rather than a cardiac hormone.

Among the natriuretic peptide receptors, it was elucidated that ANPR-B is selectively activated by CNP and considered to be the specific receptor for CNP [10,11].

124

Interestingly, ANPR-B is distributed not only in the central nervous system but also in peripheral tissues, including the blood vessel, one of the main target organs for natriuretic peptides [10,12]. These findings suggest that CNP can be an autocrine/ paracrine regulator in the periphery, like local renin-angiotensin system (RAS) [13]. The present study, therefore, aimed at the elucidation of the production and secretion of CNP in the blood vessel and examined occurrence of "vascular natriuretic peptide system", since recent evidence supports the significance of vascular RAS for vascular tone and remodeling [14], and natriuretic peptides are demonstrated to affect both vascular tone and growth in an antagonistic way to RAS [15,16].

Secretion of CNP from cultured endothelial cells

Using specific radioimmunoassay (RIA) for CNP [6], we detected CNP-like immunoreactivity (-LI) in the culture medium conditioned with endothelial cells derived from the bovine carotid artery [17]. CNP-LI in the conditioned medium was accumulated in a time-dependent manner and reached to 17.8 ± 1.6 fmol/10^6 cells/24 hr. The endothelial cellular content of CNP-LI was 0.15 ± 0.05 fmol/10^6 cells. High performance gel permeation chromatographic analysis revealed that CNP-LI in the condition medium consisted of two components corresponding to synthetic CNP and CNP-53, an N-terminally extended form of CNP [5]. Northern blot analysis detected CNPmRNA with a size of 1.2 kb in the endothelial cells. These results demonstrate the synthesis and secretion of CNP by the cultured endothelial cells.

Regulation of endothelial secretion of CNP

Regulation of the endothelial secretion of CNP by various vasoactive substances, including growth factors and cytokines, was examined [17,18] (Fig. 1). Among them, transforming growth factor β, one of the crucial factors involved in vascular remodeling [15], induced more than two-order of magnitude increase of the

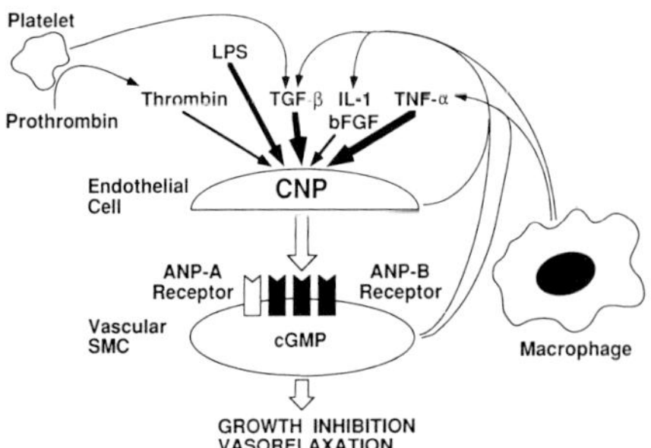

Fig. 1. Regulation of endothelial secretion of CNP and its significance in vascular endothelial cells/smooth muscle cells/macrophages interaction.

endothelial CNP secretion. In addition, the endothelial secretion of CNP is potently stimulated by tumor necrosis factor-α (TNF-α), one of the central cytokines for inflammation and also by lipopolysaccharide (LPS), the endotoxin itself.

These results suggest that CNP can act as a novel endothelium-derived relaxing factor (EDRF) in vascular endothelial cells/smooth muscle cells/macrophages interaction.

Endothelial CNP secretion in the co-culture system with vascular endothelial cells and smooth muscle cells

In order to assess the pathophysiological significance of endothelium-derived CNP, endothelial CNP secretion was examined in vascular endothelial cells/smooth muscle cells co-culture system. Co-culture of bovine vascular endothelial cells (EC) with the same number of vascular smooth muscle cells (SMC) derived from the rat aorta augmented endothelial CNP secretion by 58 fold (11.5 ± 0.55 fmol/dish/48 hr EC alone vs. 671 ± 119 fmol/dish/48 hr, EC + SMC co-culture). This finding suggests that endothelial CNP secretion is regulated in the interaction of vascular endothelial cells and smooth muscle cells in the blood vessel.

Demonstration of vascular natriuretic peptide system in vivo

We examined whether CNP gene expression occurs in the in vivo blood vessel wall by reverse transcription and polymerase chain reaction (RT-PCR) [19,20]. Our RT-

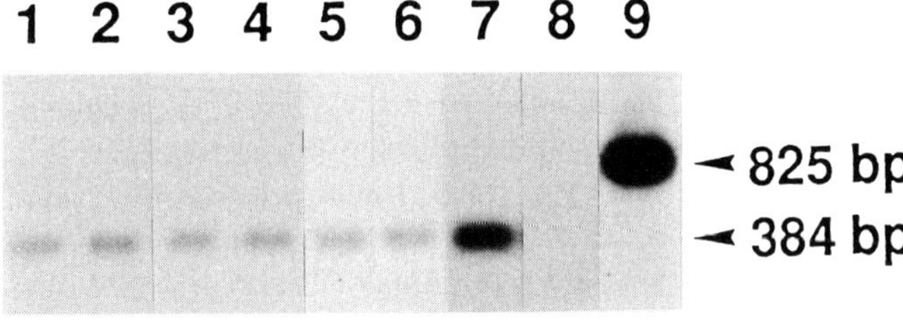

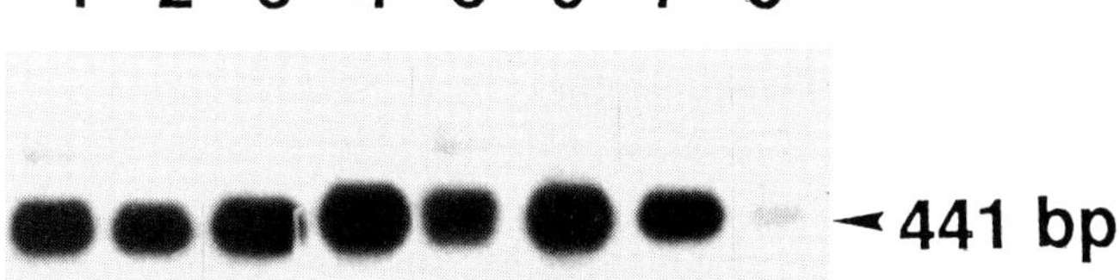

Fig. 2. Detection of gene transcripts of CNP (upper panel) and ANPR-B (lower panel) in in vivo human arteries by reverse transcription-polymerase chain reaction. Lane 1–4: internal thoracic artery; lane 5,6: gastro-epiploic artery; lane 7: brain; lane 8: liver; lane 9: genomic DNA.

PCR revealed that CNP gene transcripts were detected in blood vessels in vivo of a wide range of mammals including not only the bovine but also human, rabbit, rat and hamster aortae. Furthermore, as shown in Fig. 2, the gene transcript for ANPR-B, the specific receptor for CNP [10], was simultaneously detected with that for CNP in human arteries at various sites. Detection of CNP transcript with its receptor in in vivo vascular wall strongly indicates the presence of the "vascular natriuretic peptide system", in which CNP serves as a local regulator of vascular tone and growth.

Gene regulation of vascular natriuretic peptide system in vascular remodeling

The significance of vascular natriuretic peptide system in vascular remodeling was examined by assessing intimal-thickening lesions after carotid injury in rats. A Fogarty catheter was introduced into the left carotid artery, inflated, and withdrawn to produce endothelial denudation in male SD rats. The mRNA expression of CNP and ANPR-B was determined by the quantitative RT-PCR method, using injured and contralateral uninjured carotid arteries 2 weeks after injury. The gene transcript for CNP was detectable both in uninjured and injured blood vessels. The mRNA expression of ANPR-B was significantly augmented in the injured carotid arteries, which is in accordance with our previous report on the up-regulation of ANPR-B gene in VSMC with synthetic phenotype [21].

Detection of plasma CNP levels in humans and rats

We examined the presence of CNP in circulation in humans and rats [22]. The dilution curves of the plasma extracts from both humans and rats were parallel to the standard curve of CNP in the RIA. The plasma CNP-LI level of CNP in healthy persons was 1.4 ± 0.6 fmol/ml ($n = 13$) [22] and that of male Wistar rats (200–300 g) was 1.2 ± 0.2 fmol/ml ($n = 3$). While there was no alteration in the plasma CNP-LI level in patients with congestive heart failure or hypertension, the plasma level of CNP was markedly increased in patients with septic shock (13.2 ± 10.1 fmol/ml, $n = 11$) [22]. TNF-α is recognized to be the central mediator of the septic shock and its production is enhanced in the syndrome. The increased plasma level of CNP in sepsis is, therefore, the consequence of augmented CNP secretion from vascular endothelial cells, since TNF-α is one of the potent stimulator of endothelial CNP secretion [18]. It is also conceivable that the circulating level of CNP can be, at least in part, originated from vascular endothelial cells.

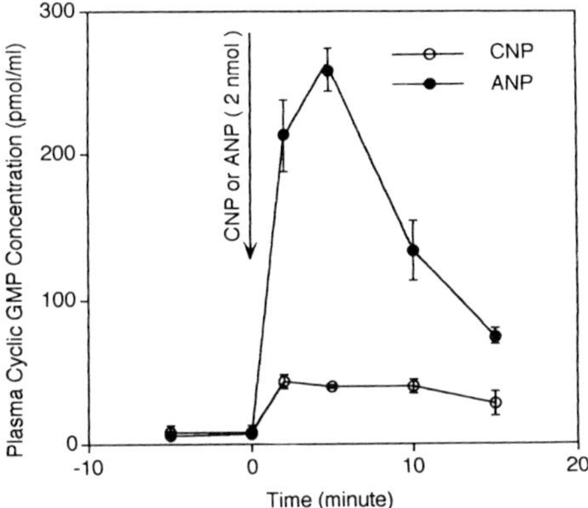

Fig. 3. Effect of the intravenous administration of CNP or ANP into the conscious rat on the plasma cyclic GMP level.

Effect of exogenously administered CNP in humans and rats

To examine whether CNP can elicit biological actions in periphery, we investigated the effect of synthetic CNP administered intravenously in humans and rats and compared with that of ANP. As shown in Fig. 3, intravenous injection of CNP (2 nmol) into the conscious rat increased the plasma level of cyclic GMP, which is the intracellular second messenger for natriuretic peptides, from 8.8 ± 1.6 pmol/ml to 39 ± 2.2 pmol/ml 5 minutes after CNP injection [23]. In contrast, the intravenous administration of ANP at the same dose increased the plasma cyclic GMP level up to 258 ± 15 pmol/ml 5 minutes after the injection.

In humans, the intravenous administration of CNP (0.43 nmol/kg) into healthy volunteers also caused 1.8-fold increase of the plasma cyclic GMP level accompanied by significant hypotensive effect and diuretic and natriuretic action. These results indicated that CNP can exert biological activities in peripheral tissues including the blood vessels in vivo.

Conclusion

Our results all together support the existence of "vascular natriuretic peptide system" in which CNP can induce vasorelaxation and growth-inhibition of vascular smooth muscle cells as an autocrine/paracrine factor. CNP can be, therefore, considered to be an endothelium-derived relaxing peptide to act as a novel mediator in the interaction of vascular endothelial cells/smooth muscle cells/macrophages in the blood vessel wall. As shown in Fig. 4, our experimental data support the speculation that in vascular proliferative lesions, the gene expressions of CNP and ANP-B receptor are up-regulated. Thus, "vascular natriuretic peptide system" could be activated as a deterrent mechanism against vascular hyperplasia/hypertrophy. The pathophysiological relevance of CNP is now under investigation in our laboratory.

128

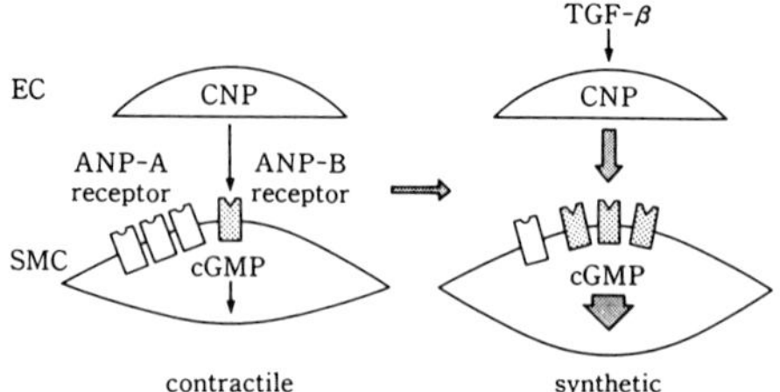

Fig. 4. Possible pathophysiological significance of vascular natriuretic peptide system in vascular remodeling.

Acknowledgements

We gratefully thank Ms K. Kito, Ms A. Takagoshi, Ms M. Shida, Ms K. Sasamoto for their secretarial and technical assistance.

This work was supported in part by research grants from the Japanese Ministry of Education, Science and Culture, the Japanese Ministry of Health and Welfare "Disorders of Adrenal Hormone" Research Committee, Japan, 1993, Smoking Research Foundation, Yamanouchi Foundation for Research on Metabolic Disorders, Salt Science Research Foundation.

References

1. Nakao K, Ogawa Y, Suga S, Imura H. J Hypertens 1992;10:907–912.
2. Nakao K, Ogawa Y, Suga S, Imura H. J Hypertens 1992;10:1111–1114.
3. Mukoyama M, Nakao K, Hosoda K, Suga S, Saito Y, Ogawa Y, Shirakami G, Jougasaki M, Obata K, Yasue H, Kambayashi Y, Inouye K, Imura H. J Clin Invest 1991;87:1402–1412.
4. Sudoh T, Minamino N, Kangawa K, Matsuo H. Biochem Biophys Res Commun 1990;168: 863–870.
5. Minamino N, Kangawa K, Matsuo H. Biochem Biophys Res Commun 1990;170:973–979.
6. Komatsu Y, Nakao K, Suga S, Ogawa Y, Mukoyama M, Arai H, Shirakami G, Hosoda K, Nakagawa O, Hama N, Kishimoto I, Imura H. Endocrinology 1991;129:1104–1106.
7. Ogawa Y, Nakao K, Nakagawa O, Komatsu Y, Hosoda K, Suga S, Arai H, Nagata K, Yoshida N, Imura H. Hypertension 1992;19:809–813.
8. Kaneko T, Shirakami G, Nakao K, Nagata I, Nakagawa O, Hama N, Suga S, Miyamoto S, Kubo H, Hirai O, Kikuchi H, Imura H. Brain Res 1993;612:104–109.
9. Shirakami G, Itoh H, Suga S, Komatsu Y, Hama N, Mori K, Nakao K. Neurosci Lett 1993;159: 25–28.
10. Suga S, Nakao K, Hosoda K, Mukoyama M, Ogawa Y, Shirakami G, Arai H, Saito Y, Kambayashi Y, Inouye K, Imura H. Endocrinology 1992;130:229–239.
11. Koller KJ, Lowe DG, Bennett GL, Minamino N, Kangawa K, Matsuo H, Goeddel DV. Science (Wash. DC) 1991;252:120–123.
12. Suga S, Nakao K, Mukoyama M, Arai H, Hosoda K, Ogawa Y, Imura H. Hypertension 1992;19: 762–765.
13. Dzau VJ. Circulation 1988;77(suppl I):I-4–I-13.
14. Itoh H, Mukoyama M, Pratt RE, Gibbons GH, Dzau VJ. J Clin Invest 1993;91:2268–2274.
15. Itoh H, Pratt RE, Dzau VJ. J Clin Invest 1990;86:1690–1697.
16. Itoh H, Pratt RE, Dzau VJ. Biochem Biophys Res Commun 1991;176:1601.

17. Suga S, Nakao K, Itoh H, Komatsu Y, Ogawa Y, Hama N, Imura H. J Clin Invest 1992;90: 1145–1149.
18. Suga S, Itoh H, Komatsu Y, Oagwa Y, Hama N, Yoshima T, Nakao K. Endocrinology 1993;133:3038–3041.
19. Komatsu Y, Nakao K, Itoh H, Suga S, Ogawa Y, Imura H. Lancet 1992;340:622.
20. Komatsu Y, Itoh H, Suga S, Ogawa Y, Arai H, Kishimoto I, Nakagawa O, Hama N, Tamura N, Takaya K, Miyamoto Y, Yoshimasa T, Matsuda K, Ikeda T, Ban T, Nakao K. Therapeutic Res 1993;14:174–183.
21. Suga S, Nakao K, Kishimoto I, Hosoda K, Mukoyama M, Arai H, Shirakami G, Ogawa Y, Komatsu Y, Nakagawa O, Hama N, Imura H. Circ Res 1992;71:34–39.
22. Hama N, Itoh H, Shirakami G, Suga S, Komatsu Y, Yoshima T, Tanaka I, Mori K, Nakao K. Biochem Biophys Res Commun 1994;198:1177–1182.
23. Hama N. A monoclonal antibody to C-type natriuretic peptide – Preparation and application to radioimmunoassay and neutralization experiment. J Endocrinol (in press).

Endothelium-Derived Factors and Vascular Functions. T. Masaki, ed.

Vasoactive factors in vascular remodeling: new insights from in vivo gene transfer technology

Gary H. Gibbons, Richard E. Pratt and Victor J. Dzau
Falk Cardiovascular Research Center, Stanford University School of Medicine, Stanford, CA, USA 94305

Summary

The vasculature is a complex, integrated organ capable of autonomous regulation of its tone and structure. Blood vessels remodel themselves in accordance with a dynamic interaction between hemodynamic, vasoactive and growth factor stimuli. The role of autocrine-paracrine vasoactive factors such as angiotensin II and nitric oxide in this process of vascular remodeling in vivo remained to be determined. In this review we discuss new insights into the role of these factors in vascular remodeling in vivo based upon a novel experimental approach using in vivo gene transfer technology.

Vascular remodeling: an emerging concept

Advances in vascular biology over the past decade have revolutionized our conception of the vasculature. It is now evident that the vessel wall is an active, integrated organ composed of endothelial, smooth muscle, and fibroblast cells coupled to each other in a complex autocrine-paracrine set of interactions. The vasculature is capable of sensing changes within its milieu, integrating these signals by intercellular communication and adapting by the local production of mediators that influence structure as well as function. Vascular remodeling is an active process of structural alteration that involves the modification of at least three basic processes: cell growth and death, cell adhesion and migration, and extracellular matrix production and degradation. These changes in vessel architecture are dependent upon a dynamic interaction between hemodynamic stimuli, vasoactive substances and locally generated growth factors. The remodeling response is usually an adaptive process in response to chronic changes in hemodynamic conditions and/or humoral factors. However, vascular remodeling may subsequently contribute to the pathophysiology of vascular diseases and circulatory disorders [1].

A spectrum of structural alterations of the blood vessel is illustrated in Fig. 1. In response to increased arterial pressure the blood vessel structure is altered such that the wall to lumen ratio is increased. The increase in wall to lumen ratio occurs by either an increase in muscle mass or rearrangements of cellular and noncellular elements (Fig. 1, vessel a) that confers an increase in vascular reactivity and potentiates the increase in peripheral resistance in hypertension [1,2]. Another form

VASCULAR REMODELING

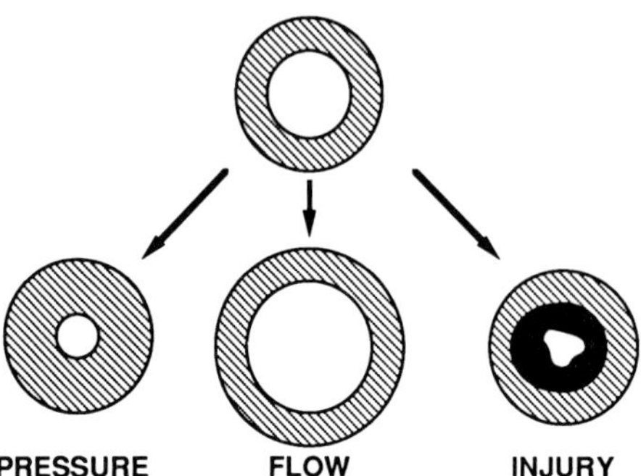

Fig. 1. Spectrum of vascular remodeling. The vascular response to increased pressure (a), increased flow (b) and vascular injury (c).

of vascular remodeling primarily involves changes in overall lumen dimensions (Fig. 1, b). In this example, active restructuring of the cellular and non-cellular components of the vessel wall results in dramatic changes in lumen dimensions with relatively small changes in medial thickness. Clinical examples of this form of remodeling include the vascular dilation associated with a sustained high blood flow state (e.g., A-V fistula) or the weakening of the vessel wall that results in aneurysm formation. Conversely, a frank reduction of the vascular mass and caliber results from a chronic reduction in blood flow. Indeed, rarefaction of the microcirculation (a loss of capillary area) is another form of vascular remodeling that promotes hypertension and tissue ischemia [3]. The architecture of the vessel wall is also markedly altered in response to vascular injury (Fig. 1, vessel c). The neointima forms as part of a reparative response involving thrombosis, vascular cell migration, proliferation, matrix production and inflammatory cell infiltration. The interplay between pressure, flow and the humoral factors activated during the reparative process determines the form and extent of neointima formation and the degree of vessel stenosis.

Although all of the cells of the blood vessel may participate in the remodeling process, the endothelium is particularly suited to play a prominent role in vascular remodeling. The endothelial surface is a dynamic interface constantly exposed to changes in the milieu produced by humoral factors, inflammatory mediators, and physical forces. The local synthesis of endogenous vasoactive substances (vaso-constrictive as well as vasodilatory factors), procoagulants as well as anticoagulants, proinflammatory as well as anti-inflammatory mediators, and growth promoters as well as growth inhibitors modulates vessel tone and structure. A constant theme of these interactions involves a delicate yin-yang balance between countervailing mechanisms such that a homeostatic balance is maintained. Thus, the endothelium is strategically located to serve as a sensory cell assessing hemodynamic and humoral signals as well as an effector cell eliciting biological responses that may eventually affect the structure of the blood vessel. In this brief review, we will examine the interaction between vascular cells in this process of vascular remodeling and discuss the role of hemodynamic stimuli and locally generated vasoactive substances as determinants of vascular structure. In particular, we will focus on new insights gained from a novel experimental approach to define the biological role of autocrine-paracrine factors utilizing in vivo gene transfer technology.

Flow-induced vascular remodeling

A large body of evidence indicates that hemodynamic stimuli such as shear stress (the tractive force exerted by flow) have profound effects on vascular function and structure [4,5]. However, the precise molecular mechanisms by which the endothelium senses hemodynamic stimuli and promotes these changes in vessel function and structure remain to be further characterized. Electrophysiologic studies indicate that the endothelium may sense changes in shear stress via the activation of K^+ channels that result in membrane hyperpolarization and changes in intracellular calcium [6–8]. For example, we have previously reported that flow-activated K^+ channels appear to play an important signal transduction role in the acute shear stress-stimulated generation of the endothelium-derived vasodilator nitric oxide (NO) [9].

In addition to acute changes in vascular tone, the endothelium appears to mediate long-term changes in vessel structure. Studies by Langille and O'Donnell [10] indicate that the endothelium plays a critical role as the blood vessel's mechano-transducer mediating the effect of these chronic changes in flow on vascular architecture. Chronic changes in blood flow also modulate the process of vascular lesion formation in diseases such as atherosclerosis, vein graft stenosis, and restenosis after angioplasty [5,11,12]. It is a longstanding yet intriguing observation that atheroma formation appears to be inhibited in areas of the circulation exposed to increased laminar shear stress [5,11]. In accord with this finding, Kraiss et al. [12] have recently shown that modest increases in shear stress within the physiologic range inhibit neointimal hyperplasia after vascular injury. It is postulated that these flow-induced changes in vessel structure are mediated via the induction of endothelium-derived mediators of remodeling. These data suggest that in response to increased shear stress, the endothelium apparently expresses a potent paracrine factor(s) that inhibits vascular smooth muscle cell proliferation in vivo. The inhibitory factor(s) that mediates this response remains to be defined.

To characterize the molecular mechanisms of flow-induced vascular remodeling, we have established an in vitro model system to expose endothelial cells to well-defined flow fields based upon a cone-plate viscometer design. In initial studies we observed that conditioned medium obtained from endothelial cells exposed to shear stress contained a stable and potent inhibitor of both endothelial cell and vascular smooth muscle cell growth. Recently, we have demonstrated that this potent inhibitor of vascular cell growth is transforming growth factor beta-1 (TGFβ1) [13]. Furthermore, we have observed that shear stress induces increased TGFβ1 mRNA expression at the transcriptional level as documented by nuclear run-on studies.

134

Moreover, blockade of flow-activated potassium channels significantly inhibits the shear stress-stimulated increase in TGFβ1 gene transcription. Recent studies indicate that there are novel gene regulatory cis elements that mediate the effect of shear stress on endothelial cell gene expression [13,14].

These findings provide the first characterization of a unique mechanotransduction pathway involving shear stress, a flow-activated potassium channel and the activation of endothelial cell gene transcription. It is anticipated that future studies will further define novel cis and trans regulatory elements mediating the effect of shear stress on gene expression regulation. In addition to these studies of TGFβ1, our laboratory and others have documented changes in the expression of several potential mediators of flow-induced remodeling including: endothelin-1, nitric oxide synthase, platelet-derived growth factor (PDGF)-BB, basic fibroblast growth factor (FGF), tissue plasminogen activator, and the leukocyte adhesion molecule VCAM-1 [4,14–16]. These studies further substantiate the concept of vascular remodeling in which a dynamic interaction between hemodynamic stimuli, vasoactive substances, adhesion molecules and growth factors promotes changes in vessel structure that modulate lesion formation. It is anticipated that further investigation in this area will provide new insights into the molecular basis of flow-induced vascular remodeling.

Role of angiotensin II in vascular remodeling

The regulation of vascular tone involves the balance between vasoconstrictors and vasodilators. A similar yin-yang balance appears to be involved in the regulation of cell growth and vessel structure. A growing body of evidence indicates that vasoactive substances have acute effects on vessel tone but may induce long-term changes in vessel structure via modulating vascular cell growth, migration or extra-cellular matrix composition. Angiotensin II (Ang II) serves as a useful paradigm of a vasoactive substance that can modulate the cellular processes of vascular remodeling. Experimental work utilizing animal models have documented that the infusion of Ang II induces vascular hypertrophy independent of its effects on blood pressure [17] and potentiates the neointimal hyperplasia that results from balloon-induced vascular injury [18]. Ang II can induce either vascular smooth muscle cell hypertrophy or hyperplasia in vitro. The cellular growth response to Ang II involves the induction of proto-oncogenes c-fos, c-jun and c-myc mRNA expressions as well as the autocrine growth factors PDGF-A chain, basic FGF and TGFβ1 [19–21]. The net growth response to Ang II, hypertrophy versus hyperplasia, is dependent on the relative balance between the effect of proliferative (PDGF, basic FGF) versus anti-proliferative (TGF-β1) autocrine growth factors and the cellular milieu. In addition to these effects on cell growth, Ang II appears to be involved in modulating angiogenesis, cell migration, and matrix production [1,22,23]. These pleiotropic properties suggest that Ang II may be an important mediator of vascular remodeling in vivo.

The potential role of Ang II in vascular remodeling in vivo is determined by the factors that regulate the quantity and localization of its expression. Studies in our laboratory and others have established that there is an autocrine-paracrine vascular angiotensin system in addition to the classical endocrine renin-angiotensin system [24]. This local vascular system appears to be activated during states of vascular remodeling. During the dramatic changes in vessel structure that occur during the development of the newborn, there is a high level expression of angiotensin converting enzyme (ACE) within smooth muscle cells throughout the media [25]. During this same period, there is high expression of the type-2 angiotensin (AT-2) receptor as well as the AT-1 receptor [26]. As the animal approaches adolescence the expression of AT-2 receptors disappears and ACE expression becomes limited to the endothelium. We speculate that the vascular angiotensin system may play a role in the ontogeny of the vasculature.

It is intriguing that in response to vascular injury, this ontogeny program becomes reactivated and these components of the vascular angiotensin system are again expressed at high levels. Our laboratory has reported increased expression of angiotensinogen, ACE and the AT-2 receptor within neointimal smooth muscle cells after vascular injury [27–29]. Analogous to the response to injury, the hemodynamic and humoral stimuli associated with hypertension also activates this local vascular angiotensin system resulting in increased local generation of Ang II [30]. Indeed, blockade of Ang II generation by ACE inhibition prevents vessel hypertrophy in hypertension [31] and inhibits neointima formation in the rat carotid balloon injury model [32]. These data provide indirect evidence that the local vascular angiotensin system may be a mediator of vascular remodeling. However, these pharmacologic studies are unable to adequately distinguish between the influence of the local autocrine-paracrine system and the classical endocrine renin-angiotensin system.

To address the role of the autocrine-paracrine vascular angiotensin system in vascular remodeling we have developed a novel experimental approach involving in vivo gene transfer technology. Utilizing a highly efficient transfection method that employs an inactivated Sendai virus coupled to DNA complexed within liposomes (HVJ method) we transfected cultured vascular smooth muscle cells with an expression vector containing the ACE gene driven by a beta-actin promoter and CMV enhancer. Our findings indicated that ACE is a rate-limiting step in the generation of Ang II by vascular cells, and that increased ACE expression is associated with Ang II-mediated vascular smooth muscle cell growth in vitro [33].

Having defined the feasibility of this approach in vitro, we employed the HVJ method to simulate the effect of increased local ACE expression within medial smooth muscle cells in vivo as observed in the newborn aorta and in hypertensive vessels. A segment of the rat carotid artery was isolated from the circulation by temporary ligatures, the expression vector containing the ACE gene within the HVJ complex was incubated in the vessel lumen for 15 minutes, and then blood flow restored by release of the ligatures. At 3 days after transfection, we observed increased ACE activity within the ACE gene-transfected vessel compared to the control vector-transfected vessel in association with increased DNA synthesis and DNA content. At 2 weeks after ACE gene transfection, there was vessel hypertrophy as evidenced by increased vessel wall protein content and an increased wall:lumen ratio. Moreover, the vascular hypertrophy induced by transfection of the ACE gene was reversed by administration of the specific AT-1 receptor antagonist Dup 753

Endothelium-Derived Factors and Vascular Functions. T. Masaki, ed.

Growth-regulatory molecules and atherosclerosis

Russell Ross

Department of Pathology, SM30, University of Washington, Seattle WA 98195, U.S.A.

Summary

The formation of lesions of atherosclerosis may result in myocardial infarction, stroke, or gangrene. The Response-to-Injury Hypothesis of Atherosclerosis proposes that such lesions exemplify an excessive, inflammatory-fibroproliferative response in the artery wall to various forms of injury. What may begin as a protective response can become a disease entity in an excessive state. Both the initial (inflammatory) and the subsequent (fibroproliferative) responses involve numerous cellular interactions among the lining endothelium, smooth muscle, monocyte-derived macrophages, and T lymphocytes, with the formation of vasoactive molecules, cytokines and growth-regulatory molecules by the cells.

Introduction

The process of atherogenesis ultimately may lead to occlusive lesions of atherosclerosis, involving both muscular and elastic arteries, and can result in myocardial infarction, stroke, gangrene and loss of function of the extremities. The lesions that form involve the intermediate and large muscular arteries of the heart, brain, and lower extremities and the elastic arteries, such as the aorta, and represent the principal cause of death in the United States, Europe, and Japan [1] .

The Response-to-Injury Hypothesis of Atherosclerosis proposes that the lesions of atherosclerosis represent an excessive, inflammatory-fibroproliferative response of the arteries to various forms of injury to the lining endothelium and underlying smooth muscle cells of the artery wall [2–5]. This response, which is meant to be a protective response, may become recurrent if the source of the injury is repetitive, chronic, or longstanding, and, thus, when it is recurrent, could become excessive. In its excess, what begins as a protective, inflammatory-fibroproliferative response may itself become a disease entity. As detailed below, the Response-to-Injury Hypothesis of Atherosclerosis suggests that, when excessive, this response can lead to the clinical sequelae of myocardial or cerebral infarction. Both the initial inflammatory response and the subsequent healing, or fibroproliferative, response involve numerous cellular interactions among the lining endothelium, smooth muscle, monocyte-derived macrophages, and T lymphocytes, with the formation of vasoactive molecules, cytokines and growth-regulatory molecules by the cells.

146

streak remains as such, or whether it progresses to an intermediate lesion and, ultimately, to an advanced lesion of atherosclerosis, termed a fibrous plaque or advanced, complicated lesion (Fig. 2).

The activated macrophages may interact with T lymphocytes by the formation and presentation of antigens to the T cells within the lesions. Both T cells and macrophages can form cytokines and growth factors that can induce migration and proliferation of each other, as well as of medial smooth muscle cells into the intima, or proliferation of pre-existing intimal smooth muscle cells. The smooth muscle cells can form new connective tissue matrix, including several forms of collagen, proteoglycans, and elastic fiber proteins. Whether or not the inflammatory process leads to the formation of a fibroproliferative response, and thus a fibrous plaque or advanced, complicated lesion of atherosclerosis, will depend in part on which genes are expressed by the macrophages, the T lymphocytes, and the smooth muscle cells, as well as by the overlying endothelium. A balance may or may not be obtained between the different growth-stimulatory and growth-inhibitory molecules and cytokines that will ultimately determine whether the lesions expand, regress, or remain stationary.

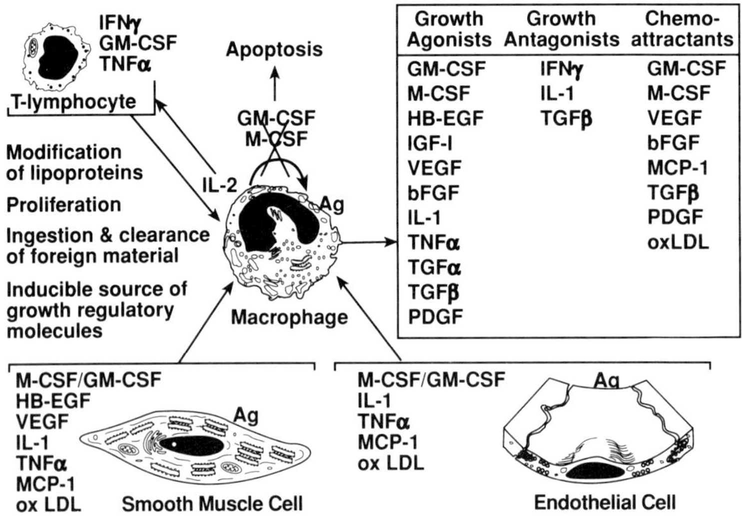

Fig. 2. The potential roles of the macrophage in atherogenesis. The reverse arrows between the T lymphocyte and macrophage suggest that some form of immune response may occur during atherogenesis. Interactions between T cells and macrophages can result in proliferation of each of these cell types through IL-2 and CSFs, respectively. All of the cells with which the macrophage can interact, namely, the T lymphocyte, smooth muscle, and endothelium, can present CSF to the macrophages to maintain cell viability and prevent apoptosis and cell death, and participate in further macrophage activation and replication. In addition, smooth muscle and endothelium can present antigens (Ag) at their surfaces and secrete chemoattractants for macrophages, including MCP-1 and oxLDL, as well as factors that can alter macrophage metabolism such as IL-1 or TNFα. When the macrophages are activated, they can produce an extraordinary number of biologically relevant molecules, some of which are listed in the box in relation to their capacity to induce or inhibit replication of endothelium, smooth muscle, or macrophages themselves, as well as their capacity to make chemoattractants for each of these cell types. Reproduced with permission from *Nature.*

It would appear that every phase of the process of atherogenesis is reversible because there is ample evidence clinically and in experimental animals to demonstrate that each of these lesions can, in fact, regress [14,15].

Growth-regulatory molecules and cytokines

Figure 2 outlines the potential interactions between the monocyte-derived macrophage in the center of the diagram and the smooth muscle cells, T lymphocytes, and endothelium and lists a host of growth-stimulatory, growth-inhibitory, and chemotactic molecules that can be generated by the activated monocyte-derived macrophages. Principal among the growth-stimulatory molecules are both chains of platelet-derived growth factor (PDGF-A, PDGF-B), insulin-like growth factor-1 (IGF-1), fibroblast growth factor (FGF), and heparin-binding EGF-like growth factor (HB-EGF). In addition, cytokines commonly formed by activated macrophages include interleukin-1 (IL-1), tumor necrosis factor α (TNFα), and the growth-regulatory molecule, transforming growth factor β (TGFβ), which is perhaps the most potent connective tissue-stimulatory molecule thus far discovered. Monocyte-derived macrophages can also form granulocyte-macrophage colony stimulating factor (GM-CSF), and monocyte colony stimulating factor (M-CSF), both of which can provide chemotactic attraction for additional monocytes, enhance their stability in the lesion, and may stimulate their replication. The latter may be important in that not only do smooth muscle cells proliferate within the lesions of atherosclerosis but monocytes/macrophages also actively multiply in the lesions as well. Other molecules that may be formed and which can be chemotactic for monocytes include monocyte chemotactic protein-1 (MCP-1) and possibly forms of modified or oxidized low-density lipoprotein (oxLDL).

Thus, the factors that may be important in determining whether the process of atherogenesis progresses or not are related to those molecules that can induce or inhibit gene expression in each of the different cells in the lesions. For example, PDGF is an extraordinarily potent mitogen and chemoattractant for smooth muscle cells [16] and is present in macrophages in lesion of atherosclerosis in humans and experimental animals [17]. It is conceivable that it may play a role in migration and chemotaxis of smooth muscle cells in the process of atherogenesis, together with several other molecules. PDGF may also be a possible common denominator resulting from cytokine stimulation because the growth-stimulatory properties of IL-1, TNFα, and TGFβ for smooth muscle cells all appear to be mediated by their capacity to induce gene expression for PDGF-BB or PDGF-AA in endothelial cells or smooth muscle cells, respectively [18,19].

Signaling pathways for chemotaxis versus proliferation

Recently, studies examining the role of PDGF-B chain versus IGF-1 in inducing smooth muscle migration versus proliferation in vitro have provided clues to two separate signaling pathways induced by these two molecules. Bornfeldt et al. [20] have demonstrated that IGF-1 and PDGF-BB are both potent inducers of smooth muscle chemotaxis. In some experiments, IGF-1 and PDGF-BB were almost equipotent in their capacity to induce chemotaxis or directed migration of smooth muscle cells in vitro. In sharp contrast, PDGF-BB is ten- to twentyfold more potent than IGF-1 as an inducer of DNA synthesis and smooth muscle proliferation in vitro. To take advantage of these observations, Bornfeldt et al. [20] went on to show that IGF-1 and PDGF-BB have similar potencies in their capacity to induce phospholipase activation, PIP_2 formation and inositol tris phosphate (IP_3) formation, and are equipotent in inducing calcium mobilization in human arterial smooth muscle cells. This equality in activation of the inositol phosphate pathway and calcium mobilization correlates well with the near equipotent capacity of these two factors to induce smooth muscle cell chemotaxis.

In contrast, after inducing phosphorylation of its receptor, PDGF-BB is a potent inducer of the mitogen-activated protein (MAP) kinase system, inducing activation of MAP kinase kinase and MAP kinase, as well as several other cytosolic factors associated with phosphorylation of a series of intermediate signaling molecules, whereas IGF-1, which is a poor mitogen, has no effect on MAP kinase kinase or MAP kinase activation [20]. These observations support the suggestion that inositol phosphate formation and calcium mobilization may be key components in the signaling pathways associated with chemotaxis of smooth muscle cells, but not with their proliferation; whereas activation of the MAP kinase cascade and its associated pathways may be key in involving smooth muscle proliferation but not their directed migration.

In summary, these studies of the process of atherogenesis and the interactions of endothelium, monocyte-derived macrophages, T lymphocytes, and smooth muscle cells have yielded insight into the genes that are expressed in each of these cell types, the signal-inducing molecules that may be formed as a result of the interactions among these cells, and the complex nature of the resultant cellular network, all of which can lead to the development of the lesions of atherosclerosis. Taken together, these set the stage for the development of new methods to both treat and prevent the process of atherogenesis.

Acknowledgements

This work is supported in part by National Heart, Lung, and Blood Institute grant HL-18645 and an unrestricted grant for cardiovascular research from Bristol-Myers Squibb Company.

References

1. Report of the Working Group of Arteriosclerosis of the National Heart, Lung, and Blood Institute. Vol. 2. Washington DC:Government Printing Office, 1981.
2. Ross R, Glomset JA. N Engl J Med 1976;295:369–377, 420–425.
3. Ross R. Arteriosclerosis 1981;1:293–311.
4. Ross R. N Engl J Med 1986;314:488–500.
5. Ross R. Nature 1993;362:801–809.
6. Virchow R. In: Gesammelte Adhandlungen zur Wissenschaftlichen Medicin. Frankfurt-am-Main:Meidinger Sohn and Company, 1856;458–636.
7. French JE. Int Rev Exp Pathol 1966;5:253–353.
8. Ross R, Glomset JA. Science 1973;180:1332–1339.
9. Hynes RO. Cell 1992;69:11–25.
10. Cybulsky MI, Gimbrone MA Jr. Science 1991;251:788–791.
11. Simionescu N, Vasile E, Lupu F, Popescu G, Simionescu M. Am J Pathol 1986;123:109–125.
12. Henriksen T, Mahoney EM, Steinberg D. Proc Natl Acad Sci USA 1981;78:6499–6503.
13. Vlassara H, Brownlee M, Cerami A. Proc Natl Acad Sci USA 1985;82:5588–5592.
14. Armstrong ML, Warner ED, Conner WE. Circ Res 1970;27:59–67.
15. Brown BG, Albers JJ, Fisher LD, Schaefer SM, Lin J-T, Kaplan C, Zhao X-Q, Bisson BD, Fitzpatrick VF, Dodge HT. N Engl J Med 1990;323:1289–1298.
16. Raines EW, Ross R. J Cell Biol 1992;116:533–543.
17. Ross R, Masuda J, Raines EW, Gown AM, Katsuda S, Sasahara M, Malden LT, Masuko H, Sato H. Science 1990;248:1009–1012.
18. Raines EW, Dower SK, Ross R. Science 1989;243:393–396.
19. Battegay EJ, Raines EW, Seifert RA, Bowen-Pope DF, Ross R. Cell 1990;63:515–524.
20. Bornfeldt KE, Raines EW, Nakano T, Graves LM, Krebs EG, Ross R. J Clin Invest (in press).

Endothelium-Derived Factors and Vascular Functions. T. Masaki, ed.

TGF-β: structure and functional consequences

Daniel B. Rifkin, Christine N. Metz, Irene Nunes and John Harpel
Department of Cell Biology and Kaplan Cancer Center, New York University Medical Center, and the Raymond and Beverly Sackler Laboratory, New York, NY 10016, USA

Transforming growth factor-β (TGF-β) is a ubiquitus protein originally identified in the conditioned media of certain tumor cells (for reviews see [1,2]). The active 25 kD molecule is a homodimer derived from a proprotein consisting of two 390 amino acid proteins. The precursor is cleaved between two arginine residues at positions 278 and 279. The mature growth factor (TGF-β), therefore, represents the carboxy terminus of the precursor molecule. In most cells and tissues the majority of the TGF-β occurs as a high molecular weight complex in which the propeptide remains associated with the mature growth factor via non-covalent interactions [1,3,4]. In this state, the TGF-β is latent and cannot interact with its receptor. Latency results only from the interaction of the propeptide with mature TGF-β as latent TGF-β can be formed by incubation of purified TGF-β and pure propeptide. The propeptide is often referred to as the latency associated protein (LAP). The latent TGF-β complex also contains an additional protein, derived from a separate gene, called the latent TGF-β binding protein (LTBP) [3,5,6]. The LTBP is covalently bound to the propeptide by a disulfide bond creating a high molecular weight form of latent TGF-β of approximately 210,000 kD. The size of the LTBP varies depending upon the cell type studied. The reason for this is not known but may represent either unique transcription products or proteolytic degradation. Thus, the latent TGF-β complex may have a variable molecular weight depending upon the cell of origin.

Five forms of TGF-β have been identified in vertebrates (TGF-β1-5) with TGF-β1-3 occurring in mammals [1,2]. These three forms of TGF-β are very similar, and a high degree of homology exists both between the different forms of TGF-β and within the same form in different species. The TGF-βs are members of a superfamily that contains many other growth factors or cytokines including activins, inhibins, the bone morphogenic proteins, Mullerian inhibitory substance and several cell surface recognition proteins [1,2].

Almost all cells that have been examined synthesize latent TGF-β. Active TGF-β interacts primarily with three different surface proteins. One of these, betaglycan, is a proteoglycan [7,8], whereas the other two are thought to be serine, threonine kinases [9]. The latter two proteins have been cloned and are believed to be the signal transducing receptor(s). The matrix proteoglycan decorin also will bind TGF-β and has been used therapeutically to neutralize TGF-β activity [10].

TGF-β was originally described as a promoter of cell growth in soft agar, but many experiments have demonstrated that it is bifunctional and can be either a growth promoter or a growth inhibitor [1,2]. The cellular response to TGF-β may vary with growth factor concentration and/or cell type. TGF-β tends to inhibit the growth of lymphoid, epithelial and endothelial cells. It is a strong inducer of the synthesis of matrix proteins such as collagen, fibronectin, and certain proteoglycans. TGF-β also is an inducer of enhanced expression of the inhibitors for plasminogen activator, plasminogen activator inhibitor-1 (PAI-1), and metalloproteinases. Therefore, in most cells exposure to TGF blocks the expression of plasminogen activator (PA) and metalloproteases. TGF-β is a potent chemoattractant for monocytes but inhibits the migration of endothelial cells, smooth muscle cells, and fibroblasts.

An important question for understanding TGF-β action is how latent TGF-β is converted into the active form. Our laboratory and the laboratory of P. D'Amore reported that cocultures of bovine endothelial cells and smooth muscle cells or pericytes converted latent TGF-β to TGF-β [11–13]. We have initiated experiments to analyze the biochemical features of this activation. The following description represents our current model for the mechanism of latent TGF-β activation.

Latent TGF-β activation generally requires the physical contact of two different cell types from the same species. Thus, TGF-β is only formed when endothelial cells and smooth muscle cells are cocultured [11,13], although both smooth muscle and endothelial cells produce latent TGF-β constitutively (Table 1). A species requirement appears to exist because bovine endothelial cells do not activate latent TGF-β in the presence of either human or pig smooth muscle cells [11]. TGF-β formation also depends upon cell-cell contact. Therefore, if endothelial cells are cultivated on a surface 1–2 mm above a monolayer of smooth muscle cells, the cells fail to produce active TGF-β. However, when smooth muscle cells and endothelial cells are co-cultured on a surface 1–2 mm above a monolayer of endothelial cells, TGF-β is produced [11,13].

The activation of latent TGF-β requires at least two proteases [11,12]. The inclusion of inhibitors of either PA or plasmin in the cocultures blocks TGF-β formation (Table 1). This observation is consistent with earlier work that demonstrated that plasmin can activate semi-purified latent TGF-β [14]. Endothelial cells and

Table 1. Latent TGF-β activation in cocultures of EC and SMC

Addition	TGF-β (pg/ml)
EC	N.D.
SMC	N.D.
EC + SMC	20
EC + SMC + aprotinin	> 5
EC + SMC + αuPA	> 5

Cocultures were established as described by Sato and Rifkin [11]. TGF-β levels were calculated by reference to a standard curve using recombinant TGF-β. EC, endothelial cells; SMC, smooth muscle cells; αuPA, antibovine urokinase plasminogen activator IgG; N.D., none detected.

smooth muscle cell both produce urokinase-type PA (uPA), and there is no apparent change in the levels of synthesis of uPA upon the initiation of coculture. Plasmin formation requires uPA to be bound to a plasma membrane binding protein, because cells deficient in the uPA receptor inefficiently activate latent TGF-β [15]. The plasmin utilized for latent TGF-β activation derives from the serum used for the initial culturing of the cells. The removal of plasminogen from the serum before utilization for cell plating abolishes TGF-β formation in cocultures [16].

Not only are the enzymes uPA and plasmin surface bound, but the substrate, latent TGF-β, also is focalized at the cell surface. This increases the local concentration of the substrate and promotes the enzyme-substrate interactions required for activation. Several mechanisms concentrate latent TGF-β at the cell surface. The first of these to be described is the interaction of mannose-6-phosphate (M6P) residues in the oligosaccharide side chains of the LAP with the plasma membrane cation-independent M6P/IGF-II receptor [16] (Table 2). Thus, the establishment of cocultures in the presence of either M6P or antibodies to the cation-independent M6P/IGF-II receptor prevents the conversion of latent TGF-β to TGF-β [17].

A second reaction required for latent TGF-β activation and that concentrates latent TGF-β at the cell surface is the interaction of LTBP with a pericellular molecule [18]. LTBP has multiple EGF-like repeats, motifs demonstrated to mediate other protein-protein interactions. This fact suggested that LTBP might function in protein-protein recognition. This was tested by preparing cocultures in the presence of antibodies to the LTBP or in the presence of excess LTBP. Under these conditions activation of latent TGF-β does not occur (Table 2). The addition of denatured LTBP to co-cultures does not block activation (Table 2). The LTBP does not appear to block the activity of plasmin or the release of latent TGF-β. Recent unpublished experiments indicate that the carboxyl-terminal end of LTBP is required for activation.

The third mechanism for concentrating latent TGF-β at the surface is via the action of the enzyme transglutaminase [19] which cross-links lysine residues to glutamine residues. Establishment of endothelial cell-smooth muscle cell cocultures

Table 2. Demonstration of a surface localization for latent TGF-β activation

Addition	TGF-β formed (pg/ml)
None	30
M6P (100 μM)	< 5
LTBP (500 ng/ml)	< 2
αLTBP (200 μg/ml)	< 2
boiled LTBP (500 ng/ml)	33
αEC TGase (10 μg/ml)	< 3
Dansylcadaverine (100 μM)	< 3

Cocultures were established as described by Sato and Rifkin [11]. Data taken from the papers of Dennis and Rifkin [17], Flaumenhaft et al. [23], and Kojima et al. [19]. αEC TGase, anti-bovine endothelial cell type II transglutaminase IgG; αLTBP, anti-latent TGF-β binding protein IgG.

with the inclusion of either inhibitors of transglutaminase or an antibody specific for bovine endothelial cell tissue transglutaminase type II prevents the formation of TGF-β (Table 2). It is possible that plasmin may be a substrate for transglutaminase in the latent TGF-β activation reaction as it is a substrate in solution [20]. Alternatively, LAP or LTBP may be a substrate for the transglutaminase.

These three features of the activation reaction undoubtedly accelerate TGF-β formation in a manner similar to the way that the localization of activated clotting factors and their substrates to surfaces enhances fibrin formation by raising the local concentrations of the reactants. In fact the activation of latent TGF-β has many similarities to the activation of clotting zymogens [18].

Although latent TGF-β is activated only under normal conditions in cocultures, cells in homotypic cultures can be induced to form TGF-β. This requires the application of specific agents. One of these agents is retinoids, which have been shown to induce TGF-β formation in cultures of either keratinocytes or endothelial cells [21,22]. Interestingly retinoid treatment induces increased levels of PA and transglutaminase in endothelial cells [22]. The requirements for latent TGF-β activation by retinoids are identical as those described for cocultures except that only one cell type must be present.

A second molecule that stimulates latent TGF-β activation in cultures of endothelial cell is basic fibroblast growth factor (bFGF) [23]. When endothelial cells are exposed to bFGF for 1 hr and the bFGF removed, this results in an increase in PA levels, a normal response to bFGF. After approximately 18 hr, however, a decrease of PA levels occurs which continues over the next 6–18 hr. The decrease of PA results from TGF-β formation. By measuring the levels of TGF-β in the culture, this can be shown directly [23]. An indirect demonstration of this effect can be shown by the inclusion of neutralizing antibodies to TGF-β in the culture after removal of the bFGF [23]. This prevents the loss of PA expression. The inclusion of antibodies to TGF-β yields PA levels significantly higher than those seen with bFGF exposure by itself. The formation of TGF-β upon bFGF exposure may illustrate a natural system for the modulation of the effects of bFGF. Application of bFGF to cells leads to activation of latent TGF-β. The TGF-β formed tends to dampen the response of endothelial cells to bFGF as TGF-β inhibits endothelial cell division, migration, and PA expression, whereas bFGF stimulates these properties. Other agents that increase PA levels in endothelial cells, such as PMA, also activate latent TBP. However, raising the PA levels is not sufficient by itself to form TGF-β.

The regulation of TGF-β production in cocultures appears to be self-regulating. This was predicted by Lyons et al. [14], who postulated that if TGF-β was formed by the PA/plasmin system, the TGF-β would induce high levels of expression of PAI-1 and that this would inhibit the activation reaction. Sato et al. [12] demonstrated this to be the case in cocultures (Table 3). Thus, the conversion of latent TGF-β to TGF-β was complete by 6–12 hr after establishment of the co-cultures. Additional TGF-β formation did not occur in the following 24 hr. However, TGF-β formation continued for periods up to 18 hr if cocultures were established in the presence of antibodies to PAI-1. Removal of the antibody resulted in the immediate

Table 3. Self-regulation of latent TGF-β activation

Condition	TGF-β (pg/ml)
Coculture 0–12 hr incubation	27
Coculture 12–24 hr incubation	< 3
Coculture 12–24 hr incubation plus α bovine PAI-1 IgG (400 µg/ml)	32

Conditions used were those reported by Sato et al. [11].

cessation of latent TGF-β activation. Moreover, PAI-1 expression increased in cocultures, and the inclusion of neutralizing antibodies to TGF-β prevented the increase. In vivo, TGF-β formation may oscilate about some value as PAI-1 and PA levels vary.

TGF-β formation can also be regulated by blocking plasminogen binding to the cell surface. Kojima et al. [16] demonstrated that the atherogenic protein Lp(a) blocks the binding of plasminogen to the cell surface in cocultures resulting in an inhibition of TGF-β formation. Because TGF-β inhibits smooth mucle cell migration [15], the inhibition of TGF-β generation by Lp(a) may contribute to the atherogenic effect of this lipoprotein.

These experiments describing the activation of latent TGF-β to TGF-β may illustrate a novel way that vascular cells interact with each other and produce mediators of their function. These interactions may have a variety of consequences for our understanding of vascular biology as well as the design of interventional therapies for certain pathological situations. The in vivo relevance of these observations in vivo awaits confirmation, however.

Acknowledgement

This research was supported by grants from the National Institutes of Health and the American Cancer Society.

References

1. Massague J. Ann Rev Cell Biol 1990;6:597–641.
2. Roberts AB, Sporn MB. In: Sporn MB, Roberts AB (eds.) Peptide growth factors and their receptors I. Berlin: Springer-Verlag, 1990;419–472.
3. Wakefield LM, Smith DM, Flanders KC, Sporn MB. J Biol Chem 1988;263:7646–7654.
4. Pircher R, Jullien J, Lawrence DA. Biochem Biophys Res Comm 1986;136:30–37.
5. Miyazono K, Hellman U, Wernstedt C, Heldin C-H. J Biol Chem 1988;263:6407–6415.
6. Kanzaki T, Olofsson A, Moren A, Werstedt C, Hellman U, Miyazono K, Claesson-Welsh L, Heldin C-H. Cell 1990;61:1051–1061.
7. Andres JL, Stanley K, Cheifetz S, Massague J. J Cell Biol 1989;109:31317–31345.
8. Wang X-F, Lin HY, Ng-Eaton E, Downward J, Lodish HF, Weinberg RA. Cell 1991;74:797–805.
9. Lin HY, Wang X-F, Ng-Eaton E, Weinberg RA, Lodish HF. Cell 1992;68:775–785.

156

10. Border WA, Noble NA, Yamamoto T, Tomooka S, Kajami S. Kidney Internat 1992;41:566–570.
11. Sato Y, Rifkin DB. J Cell Biol 1989;109:309–315.
12. Sato Y, Tsuboi R, Lyons RM, Moses H, Rifkin, DB. J Cell Biol 1990;111:757–763.
13. Antonelli-Orlidge A, Saunders KB, Smith SR, D'Amore PA. Proc Natl Acad Sci USA 1989;86: 4544–4548.
14. Lyons RM, Keski-Oja J, Moses HL. J Cell Biol 1988;106:1659–1665.
15. Odkeon L, Blasi F, Rifkin DB. J Cell Physiol (in press).
16. Kojima S, Harpel PC, Rifkin DB. J Cell Biol 1991;113:1439–1445.
17. Dennis PA, Rifkin DB. Proc Natl Acad Sci USA 1991;88:580–584.
18. Flaumenhaft R, Abe M, Sato Y, Miyazono K, Harpel JG, Heldin C-H, Rifkin DB. J Cell Biol 1993;120:995–1002.
19. Kojima S, Nara K, Rifkin DB. J Cell Biol 1993 (in press).
20. Bendixen E, Borth W, Harpel PC. Blood 1991;78:283a.
21. Glick AB, Flanders KC, Danielpour D, Yuspa SH, Sporn MB. Cell Reg 1989;1:87–97.
22. Kojima S, Rifkin DB. J Cell Physiol 1993;155:323–332.
23. Flaumenhaft R, Abe M, Mignatti P, Rifkin DB. J Cell Biol 1992;118:901–909.

Endothelium-Derived Factors and Vascular Functions. T. Masaki, ed.

Transforming growth factor-β receptors

Kohei Miyazono[1,2] and Fumimaro Takaku[2]
[1]Ludwig Institute for Cancer Research, Box 595 Biomedical Center, S-751 24 Uppsala, Sweden;
[2]International Medical Center of Japan, 1-21-1 Toyama, Shinjuku-ku, Tokyo 162, Japan

Summary

Transforming growth factor-β (TGF-β) is a family of 25 kDa dimeric proteins that have pleiotropic effects on many different cell types. TGF-β inhibits the growth and migration of endothelial cells in vitro, but it stimulates angiogenesis in vivo. TGF-β exerts its effects through binding to specific cell surface receptors. The type I (53 kDa) and type II (80 kDa) receptors, which have serine/threonine kinase domains in their cytoplasmic portions, are most important for signal transduction of TGF-β. Signal transduction by TGF-β involves the formation of heteromeric kinase receptor complexes. The type III receptor (300 kDa) and endoglin (180 kDa) are indirectly involved in the TGF-β signaling. Endoglin is a major glycoprotein of endothelial cells, which may play roles in presenting the ligands to the signaling receptors.

Introduction

Angiogenesis is an important process, in which new blood vessel is formed. Several growth factors and other bioactive molecules have been shown to induce angiogenesis in vivo (reviewed in [1]), e.g. fibroblast growth factors (FGFs), vascular endothelial growth factor (VEGF), and platelet-derived endothelial cell growth factor (PD-ECGF), which is identical to thymidine phosphorylase [2,3]. Transforming growth factor-β (TGF-β) is a family of 25 kDa dimeric proteins that have pleiotropic effects on many cell types (reviewed in [4,5]). Three different mammalian isoforms (TGF-β1, -β2, and -β3) have been identified. TGF-βs belong to a larger superfamily (TGF-β superfamily) [5], which includes activins, inhibins, bone morphogenetic proteins (BMPs), Müllerian inhibiting substance (MIS) and glial cell line-derived neurotrophic factor (GDNF). TGF-βs are produced by megakaryocytes, smooth muscle cells and endothelial cells, and they inhibit the growth and migration of endothelial cells. TGF-β is produced and secreted as latent forms [6], which need to be activated in order to bind to cell surface receptors. Although physiological mechanism for the activation of latent TGF-β is not fully understood, co-culture of endothelial cells and smooth muscle cells was shown to result in the efficient activation of latent TGF-β, which then inhibited the growth and migration of endothelial cells [7]. On the other hand, exogenously administered TGF-β stimulates angiogenesis in vivo, which is due to the migration of fibroblasts and epithelial cells and accumulation of extracellular matrix proteins [8,9]. Thus, TGF-β has been suggested to play important roles in the regulation of growth and

158

metabolism of vascular cells in vivo. In this review, we describe the structures and functions of TGF-β receptors.

TGF-β receptor system

TGF-βs exert their effects through binding to specific cell surface receptors. Four types of TGF-β receptors have been molecularly cloned and characterized, i.e. receptors type I, type II, and type III (or betaglycan), and endoglin (reviewed in [10–12]). The TGF-β type I and type II receptors are 53 kDa and 80 kDa glycoproteins, respectively, and are most important for signal transduction of TGF-β. On the other hand, the type III receptor and endoglin are indirectly involved in the TGF-β signaling. The type III receptor is a 300 kDa membrane proteoglycan [13–15], which is broadly expressed in many cells, except for certain hematopoietic cells and endothelial cells. TGF-β binds to the core protein (120 kDa) of the type III receptor, but not to the glycosaminoglycan chains. Endoglin is a 180 kDa dimeric protein, which is expressed in endothelial cells, mesangium of the kidney, and certain hematopoietic cells [16]. The transmembrane and intracellular domains of endoglin are structurally very similar to the corresponding regions of the type III receptor. Properties of the TGF-β receptors are summarized in Table 1.

Cloning of novel serine/threonine kinase receptors

The TGF-β type II receptor cDNA was obtained by an expression cloning strategy [17]. Similar to the type II receptor for activin [18], the TGF-β type II receptor has a serine/threonine kinase domain in its cytoplasmic portion. Another form of activin type II receptor (activin type IIB receptor) [19,20] has also a similar structure with the activin type II receptor. These results suggest that serine/threonine kinase receptors form a new receptor family, which may include the receptors for the proteins in the TGF-β superfamily.

The kinase domains of the serine/threonine kinase receptors show significant sequence similarity to each other. The extracellular domains of these receptors are rich in cysteine residues, but have little sequence similarity. To identify additional

Table 1. Receptors for TGF-β

	Sizes	Distribution	Binding affinities	Other properties
Type I	53kDa	most cells	TGF-β1=β3>β2	Ser/Thr kinase receptor
Type II	80 kDa	most cells	TGF-β1=β3>β2	Ser/Thr kinase receptor
Type III (Betaglycan)	300 kDa	broad	TGF-β1=β2=β3	Proteoglycan (120 kDa core protein)
Endoglin	180 kDa (dimer)	limited (e.g. EC[a])	TGF-β1=β3>>β2	Sequence similarity in part with the type III receptor

[a]EC, endothelial cells.

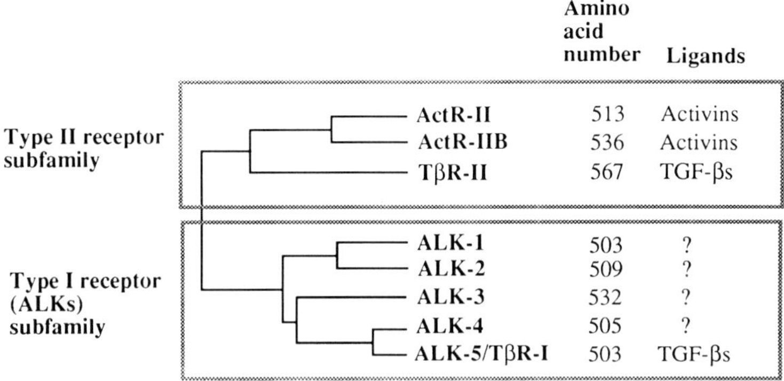

Fig. 1. Comparison of the mammalian serine/threonine kinase receptors. TβR, TGF-β receptor; ActR, activin receptor.

Table 2. Expression of ALK-1 to -5 in human tissues

Designations	ALK-1	ALK-2	ALK-3	ALK-4	ALK-5
Transcript size (kb)	2.2 (4.9)	4.0	4.4 (7.9)	5.2	5.5
Expression					
Heart	+	++	+	+	+
Brain	–	+	–	++	+
Placenta	+++	+++	+	+	+++
Lung	+++	+	–	++	++
Liver	–	+	–	++	+
Skeletal muscles	++	+++	+++	+	++
Kidney	+	++	+	+++	++
Pancreas	–	+	+	+++	+

members in the serine/threonine kinase receptor family, we have used a PCR-based approach. The strategy was based upon making use of the sequence identity in the kinase domains of the activin type II receptor and Daf-1 (serine/threonine kinase receptor in nematodes) [21], which resulted in the isolation of five novel serine/ threonine kinase receptors, denoted activin receptor-like kinase (ALK) -1, -2, -3, -4, and -5 [22,23]. The ALKs have similar domain structures, comprising of signal sequences at the N-terminals, followed by cysteine-rich, putative ligand binding domains, hydrophobic transmembrane regions and C-terminal intracellular portions which contain serine/threonine kinase domains. The kinase domains of ALKs have ~ 40% sequence identity to the type II receptors for TGF-β and activin. However, the sequence identities are higher (60–79%) among ALKs, suggesting that they form a subfamily among the serine/threonine kinase receptors (Fig. 1). The expression patterns of ALK mRNAs in human tissues are different between the different ALKs (Table 2). ALK-2, -4 and -5 are widely expressed, whereas ALK-1 and -3 have more restricted expression patterns [22,23]. These results suggest that each ALK may have different functions in vivo.

Identification of a TGF-β type I receptor

The TGF-β type I and type II receptors are indispensable for signal transduction [24,25]. The type II receptor is needed for the binding of TGF-β to the type I receptor, and the type I receptor is needed for the signal transduction induced by the type II receptor [26]. Therefore, signal transduction by TGF-β involves the formation of a heteromeric complex of two receptor molecules.

The TGF-β type I receptor has been shown to be a 53 kDa glycoprotein by affinity cross-linking studies. We have recently found that ALK-5 is a type I receptor for TGF-β [23]. ^{125}I-TGF-β1 bound to porcine aortic endothelial (PAE) cells transfected with the ALK-5 cDNA, and formed a cross-linked TGF-β type I receptor complex of 70 kDa. When the cross-linked complexes of the transfected cells were immunoprecipitated by the ALK-5 antibodies, the 70 kDa as well as the 94 kDa TGF-β type II receptor complex could be precipitated, indicating that ALK-5 formed a heteromeric complex with the type II receptor. The ALK-5 had biochemical properties of the TGF-β type I receptor; e.g. binding of TGF-β was abolished by dithiothreitol treatment of the cell surface. Moreover, the type I receptor cross-linked complexes in TGF-β responsive cells, such as mink lung epithelial cells (Mv1Lu) and human foreskin fibroblasts, could be precipitated by the ALK-5 antiserum. The structures of the type I receptor (ALK-5) and the type II receptor are shown in Fig. 2. We also found that ALK-5 is the functional type I receptor for TGF-β; transfection of the ALK-5 cDNA into TGF-β type I receptor deficient cells restored the responsiveness to TGF-β, and the cells produced plasminogen activator inhibitor-1 (PAI-1) upon addition of TGF-β. These results suggest that signal transduction by TGF-β involves the formation of a heteromeric complex of two different serine/threonine kinase receptors.

Based upon the close sequence similarity, the other ALKs may serve as type I receptors for other members in the TGF-β superfamily, i.e. activins and inhibins, BMPs, MIS, and GDNF. Identification of the ligands for the other ALKs are under way in our laboratory.

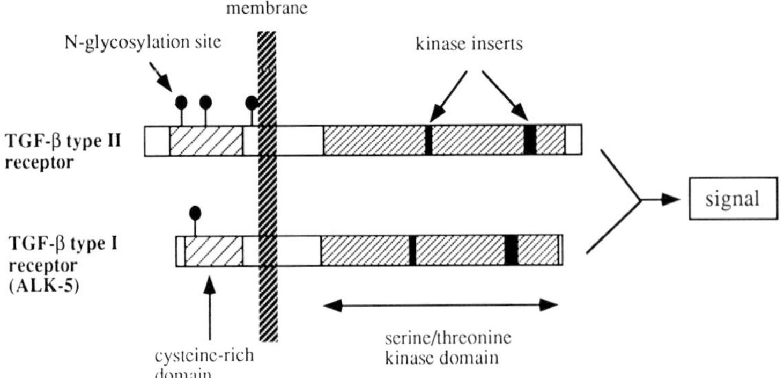

Fig. 2. Schematic illustration of the structures of the TGF-β type I and type II receptors.

Role of endoglin in the TGF-β receptor system

The TGF-β type III receptor is a membrane proteoglycan lacking cytoplasmic protein kinase domain [13–15]. The transmembrane and intracellular parts of the type III receptor are very similar (63% amino acid sequence identity) to the corresponding regions of endoglin (Fig. 3). Recently, human endoglin was shown to bind TGF-β [27,28], which suggests that endoglin has a similar function as the TGF-β type III receptor.

We have obtained a porcine cDNA clone for endoglin from a porcine uterus cDNA library [28]. The primary translated product of endoglin consists of 643 amino acids with a high sequence identity (96%) to human endoglin in the transmembrane and intracellular domains, but with a lower sequence identity (66%) in the extracellular domain. Human endoglin has an Arg-Gly-Asp (RGD) tripeptide sequence in the extracellular domain [16], which may be a possible binding site for cell adhesion proteins, but the RGD tripeptide sequence is not conserved in porcine [28]. Similar to its human counterpart, endoglin in PAE cells bound TGF-β1 and -β3 efficiently, but TGF-β2 less efficiently. Porcine endoglin was found to be co-immunoprecipitated with TGF-β type I and/or type II receptors by the endoglin antibodies or by the TGF-β type II receptor antibodies in the presence of the ligand [28]. Thus, endoglin appeared to form a heteromeric receptor complex with the type I and/or type II receptors, and act as a presentor of TGF-β for the signaling receptors. Similar results have recently been shown for the TGF-β type III receptor [29]. Moreover, endoglin was shown to be phosphorylated on serine residue(s), but the phosphorylation did not change after stimulation by TGF-β. Whether endoglin is phosphorylated by the type I and/or type II receptors, and whether it serves as a substrate of the serine/threonine kinase receptors, remain to be elucidated.

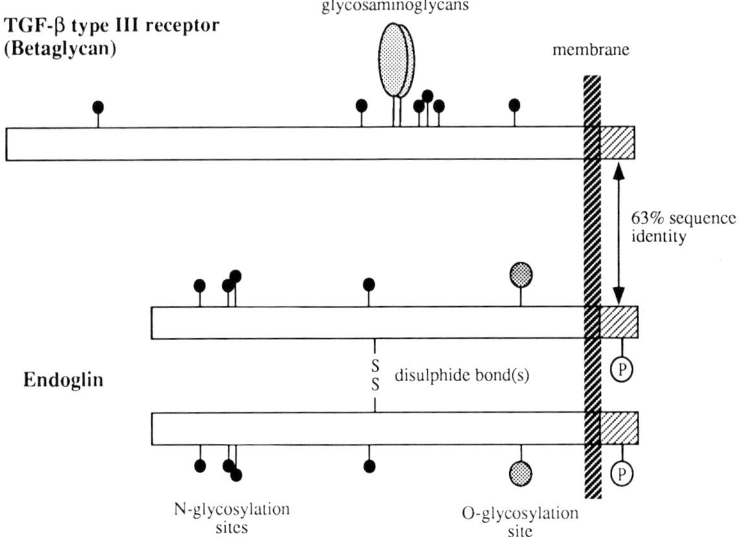

Fig. 3. Schematic illustration of the structures of human and porcine endoglin.

162

Conclusion

Most of the growth factors transduce their signals through tyrosine kinase receptors [30]. The TGF-β receptor systems are thus very different from other growth factor receptor systems. Further studies on the structures and functions of the TGF-β receptors will accelerate the elucidation of the mechanisms of TGF-β action in various physiological conditions.

Acknowledgements

We thank our collaborators Carl-Henrik Heldin, Peter ten Dijke, Hidetoshi Yamashita, Petra Franzén, Peter Schulz, Pascal Colosetti, Anders Olofsson, Fumitoshi Taketazu, Keiko Funa, Anita Morén, Susanne Grimsby, and Hidenori Ichijo at the Ludwig Institute for Cancer Research, Fuyuki Ishikawa, Hideo Toyoshima, Kensuke Usuki, and Hisamaru Hirai at the Third Department of Internal Medicine, University of Tokyo.

References

1. Folkman J, Shing Y. J Biol Chem 1992;267:10931–10934.
2. Ishikawa F, Miyazono K, Hellman U, Drexler H, Wernstedt C, Hagiwara K, Usuki K, Takaku F, Risau W, Heldin C-H. Nature 1989;338:557–562.
3. Usuki K, Saras J, Waltenberger J, Miyazono K, Pierce G, Thomason A, Heldin C-H. Biochem Biophys Res Commun 1992;184:1311–1316.
4. Roberts AB, Sporn MB. In: Sporn MB, Roberts AB (eds.) Pepide Growth Factors and Their Receptors, Part I. Berlin: Springer-Verlag, 1990;419–472.
5. Massagué J. Annu Rev Cell Biol 1990;6:597–641.
6. Miyazono K, Ichijo H, Heldin C-H. Growth Factors 1993;8:11–22.
7. Sato Y, Rifkin DB. J Cell Biol 1989;109:309–315.
8. Roberts AB, Sporn MB, Assoian RK, Smith DM, Roche NS, Wakefield LM, Heine UI, Liotta LA, Falanga V, Kehrl JH, Fauci AS. Proc Natl Acad Sci USA 1986;83:4167–4171.
9. Yang EY, Moses HL. J Cell Biol 1990;111:731–741.
10. Massagué J. Cell 1992;69:1067–1070.
11. Lin HY, Lodish HF. Trends Cell Biol 1993;3:14–19.
12. Miyazono K, ten Dijke P, Ichijo H, Heldin C-H. Adv Immunol 1994 (in press).
13. López-Casillas F, Cheifetz S, Doody J, Andres JL, Lane WS, Massagué J. Cell 1991;67:785–795.
14. Wang X-F, Lin HY, Ng-Eaton E, Downward J, Lodish HF, Weinberg RA. Cell 1991;67:797–805.
15. Morén A, Ichijo H, Miyazono K. Biochem Biophys Res Commun 1992;189:356–362.
16. Gougos A, Letarte M. J Biol Chem 1990;265:8361–8364.
17. Lin HY, Wang X-F, Ng-Eaton E, Weinberg RA, Lodish HF. Cell 1992;68:775–785.
18. Mathews LS, Vale WW. Cell 1991;65:973–982.
19. Attisano L, Wrana JL, Cheifetz S, Massagué J. Cell 1992;68:97–108.
20. Mathews LS, Vale WW, Kintner CR. Science 1992;255:1702–1705.
21. Georgi LL, Albert PS, Riddle DL. Cell 1990;61:635–645.
22. Ten Dijke P, Ichijo H, Franzén P, Schulz P, Saras J, Toyoshima H, Heldin C-H, Miyazono K. Oncogene 1993;8:2879–2887.

23. Franzén P, Ten Dijke P, Ichijo H, Yamashita H, Schulz P, Heldin C-H, Miyazono K. Cell 1993;75: 781–792.
24. Laiho M, Weis FMB, Massagué J. J Biol Chem 1990;265:18518–18524.
25. Laiho M, Weis FMB, Boyd FT, Ignotz RA, Massagué J. J Biol Chem 1991;266:9108–9112.
26. Wrana JL, Attisano L, Cárcamo J, Zentella A, Doody J, Laiho M, Wang X-F, Massagué J. Cell 1992;71:1003–1014.
27. Cheifetz S, Bellón T, Calés C, Vera S, Bernabeu C, Massagué J, Letarte M. J Biol Chem 1992; 267:19027–19030.
28. Yamashita H, Ichijo H, Grimsby S, Morén A, Ten Dijke P, Miyazono K. J Biol Chem 1994;269: 1995–2001.
29. López-Casillas F, Wrana JL, Massagué J. Cell 1993;73:1435–1444.
30. Ullrich A, Schlessinger J. Cell 1990;61:203–212.

Endothelium-Derived Factors and Vascular Functions. T. Masaki, ed.

Basic fibroblast growth factor (FGF-2) and potential mechanisms that regulate its activities

Andrew Baird

Department of Molecular and Cellular Growth Biology, The Whittier Institute for Diabetes and Endocrinology, 9894 Genesee Ave., La Jolla, CA 92037, USA

Summary

Because basic fibroblast growth factor (FGF-2) is pluripotent and nearly ubiquitous in its distribution, numerous investigators have proposed that there must exist highly specific mechanisms that regulate its bioavailability for its target cell [1,2]. Clearly, the identification of these mechanisms could serve as a first step towards developing novel strategies to inhibit FGF action and as such FGF dependent activities like angiogenesis, tumor growth, reproduction, and selected diseases of cell proliferation. Here we describe some of the processes that are thought to regulate the ability of target cells to activate the FGF-2 that is found in its local milieu.

Introduction

The fibroblast growth factor (FGF) family of growth factors consists of nine distinct, but structurally related, growth factors [3–5] that include acidic FGF (FGF-1), basic FGF (FGF-2), int-2 (FGF-3), kaposi-FGF or hst-1 (FGF-4), FGF-5, FGF-6, keratinocyte growth factor (FGF-7), androgen inducible growth factor (FGF-8) and glia activating factor (FGF-9). The superfamily also includes the cytokines IL-1α and IL-1β. Although the FGFs have numerous common characteristics, they also have significant differences. Perhaps the most salient of these is the observation that while most FGFs are potent mitogens for cells derived from the mesoderm and neuroectoderm, KGF (FGF-7) has considerable specificity for epithelial cells. All nine members of the FGF family contain a common domain in which most of the structural homology is found. There are also two of the FGFs that are distinct from other members of the family (FGF-1 and FGF-2) because their precursers do not have a signal peptide that might account for their release from cells. In spite of this observation, both FGF-1 and FGF-2 are found outside cells, are widely distributed in tissues, expressed by a large number of cells in culture and can stimulate the proliferation and differentiation of numerous cell types. Yet, to this day it is unclear how they are exported.

Despite the absence of a mechanism to account for basic FGF (FGF-2) export from cells, it appears to be translocated to the cell surface. Immunohistochemical studies show that FGF-2 localizes to the basement membrane and to the extra-

166

cellular matrix of numerous tissues, suggesting that it is sequestered outside of the target cell. If so, it is in what is presumed to be a biologically inert form because the cells surrounding this FGF are quiescent and often unresponsive to the growth factor when infused.

FGF-2 stimulates growth responses in a broad range of cell types. From a physiological perspective, it was initially not clear how a molecule which is so pleiotropic could play a physiological role in the control of cell growth and differentiation. Its pleiotropic activity was further complicated by the observation that significant amounts of this growth factor could be purified from almost any tissue in which the target cells are found. If this protein is a growth factor in vivo, how could it be present in so many tissues in which the target cells are essentially quiescent? One possibility of course is that this molecule is not a growth factor in vivo and plays some other function. Yet it is found in the extracellular matrix where it is presumably sequestered and stored in an inactive form. If that is the case, it must be highly regulated. Three components of this process may include through its receptors, by limited proteolysis or by structural modification. Each of these is described here.

Regulation of FGF activities by its low and high affinity receptor

The high affinity receptors for FGF-2 (of which there are four known genes) are tyrosine kinases that have ligand binding, extracellular domains of the IgG superfamily. Alternative splicing leads to the creation of numerous high affinity receptors from any one gene. It is generally assumed that the expression of different combinations of receptor isoforms predetermines ligand action. The low affinity proteoglycan FGF receptors also play a major role in regulating FGF action. But, it is important to note first that there are at least three types of low affinity FGF receptors: syndecans, β-glycan, and glypican. Syndecans are transmembrane receptors with heparan sulfate attached to a protein core sequence encoded by one of four genes. It is assumed that all of these proteoglycans bind FGF and that their role is to deliver FGF to high affinity receptors [6–8]. β-Glycan is also a transmembrane heparan sulfate proteoglycan which is better known for its ability to bind TGFβ. Glypican on the other hand, is an FGF-binding proteoglycan which is not transmembrane but linked to the cell surface through a phosphoinositol. In contrast, perfecan is a heparan sulfate proteoglycan that is not associated with the cell surface but that localizes to the extracellular matrix (ECM).

(a) High affinity receptor regulation of FGF activity

If the high affinity receptors for FGF are involved in the regulation of FGF-2 activities it is most likely mediated through their restricted expression in vivo. Recent studies have now shown that these receptors are present during embryonic development and that their distribution is cell type specific [9]. In normal adult tissues, there is little expression but it is readily detectable after injury, in tumors or in proliferative disease. There are certain exceptions to this rule: in the central nervous system there are focal sites of FGF synthesis that include the hippocampus, ependymal cells of the lateral and third ventricles and the subfornical organ. Generally speaking however, receptor expression is low in peripheral tissues.

If there is limited receptor expression, then FGF-2 may have a specificity and selectivity of action in spite of being both multifunctional on cells, pluripotent for different cell types and ubiquitous in its distribution. There is compelling evidence for example that cells have specific signatures of high affinity receptor isoform expression on their cell surface [9]. Increased receptor concentrations at the cell surface could also, by mass action, promote the transfer of the ligand from low affinity to high affinity sites of binding. These hypotheses are supported in part by the findings that, when FGF-2 is made cytotoxic by linking it to the ribosomal inactivating protein saporin, the mitotoxin is highly specific in vivo for proliferating cells but not to all of its presumed cell targets [10–12]. That being the case, cells can be seen as refractory to FGF unless there is an injury which activates the FGF high affinity receptor system. Similarly, numerous cells are responsive to FGF in vitro because under the conditions of cell culture, they have an injured and activated phenotype with an activated FGF receptor system.

(b) Low affinity receptor regulation of FGF activity

There are many possible mechanisms through which proteoglycan low affinity receptors might regulate FGF activity. The most likely is that they serve a dual function to sometimes sequester and other times deliver (after the appropriate signal) FGF to its high affinity receptor. The nature of this signal is not known (see below). Accordingly, the low affinity receptor complex should be viewed as another step in the regulation of FGF action. Because there are several different types of low affinity receptor protein cores and each may have multiple patterns of complexity depending on the form of heparan sulfate that is attached and the degree of sulfation, it is exceeding difficult to dissect a specific function for any one low affinity receptor. Each isoform may have different effects on FGF specificity and function; some inhibiting FGF action, others promoting it. In any event, one of the key processes in the formation of a functional ligand-receptor complex at the cell surface appears to be the transfer of ligand to the high affinity receptor [6–8].

Proteolytic regulation of FGF activity

One possible mechanism that might enable FGF-2 to leave low affinity receptors and reach the cell surface from the extracellular matrix (ECM) is through selected proteolysis. This is particularly attractive because when bound to heparin, the ligand is protected from proteolysis. When fixed sections of tissues are stained with anti-FGF antibodies [13], immunoreactive FGF is clearly associated with the ECM. There are also areas in which the immunoreactive staining is associated with the cell surface. When FGF is associated with the ECM, it can be released in a soluble form by proteolysis [14–16]. Brem et al. [17] also showed that an immunoreactive FGF-2 (when assayed by ELISA) is detected in the fluids of partial thickness and full thickness wounds suggesting that the FGF present in these fluids may have been released because of degradation of the ECM during wounding. In the retina where FGF-2 staining can be observed in the basement membrane of the retinal pigmented epithelial (RPE) cells, heparinase treatment eliminates FGF staining [18]. FGF-2 can also be released from the cell surface by the treatment of cells with phospholipase C [19] suggesting that there is an active release of the growth factor and translocation between the ECM and its tyrosine kinase receptors.

If this is correct, there should be instances in vivo where FGF is detectable in serum or biological fluids. Indeed, over the course of the last several years there have been isolated reports that have described the presence of FGF-like molecules in urine, vitreous and in blood. Most recently, the availability of a highly sensitive ELISA for FGF has resulted in the systematic analysis of FGF in cancer patients [20]. Whether this FGF is derived directly from tumor cells or after the proteolysis that accompanies tissue remodeling remains to be determined.

Regulation of FGF by post-translational modification

If the low affinity FGF receptor is not just a means of sequestering the growth factor and that it also participates in the regulation of its activity, then the question arises as to how the growth factor translocates to the high affinity from the low affinity receptor. On one hand, as described above, there could be the release of a soluble FGF that would then be available to the general circulation. Yet there are a number of instances where it has been observed that, while FGF is available to cells that have both low affinity and high affinity receptor on their surfaces, the cells do not seem to utilize their endogenous growth factor. Yet these same cells are able to respond to exogenous FGF. How then can a cell that synthesizes FGF and translocates it onto the cell surface be unable to respond to this same growth factor?

To explain this observation we hypothesized that there might exist a specific extracellular process to catalyze the growth factor's shift from low affinity receptors of sequestration towards low affinity receptors of delivery and ultimately towards high affinity receptors of signal transduction. Because this molecular event might involve a modification of the growth factor and we had observed that FGF-2 can be phosphorylated, we examined whether FGF could be phosphorylated extracellularly. We reported the detection of a non-cAMP activated protein kinase capable of phosphorylating FGF-2 on the outer cell surface of human hepatoma cells (SK-Hep cells). During the course of its characterization, we also detected a second, distinct kinase that is cAMP-independent. The addition of $[\gamma\text{-}^{32}P]$-ATP results in a rapid (< 5 min) incorporation of $[^{32}P]$ into FGF-2. The reaction is time-, ATP-, and substrate-dependent. It is unaffected by inhibitors of cAMP-dependent phosphorylation (PKI, K562b), but is modulated by calcium, magnesium, and manganese. Heparin also modulates phosphorylation on FGF-2. Gel permeation chromatography of SK-Hep cell membranes identifies two constituents of 350 and 90 kDa. From these results, we have developed our current working hypothesis that a cell surface protein kinase is a potential signal that modifies FGF-2 such that it can be translocated from extracellular sites of sequestration and recognized by its signal transducing high affinity receptor. The structural characterization of this molecule will be the ultimate test of this hypothesis and is underway.

Conclusions

In summary, the physiology of FGF regulation is complex and far from understood. It involves extracellular low and high affinity receptors, sequestered and active forms of extracellular FGF, and potentially various post-translational modifications of FGF including phosphorylation. FGF appears to be regulated at numerous steps: 1) at the level of its synthesis where gene expression (in peripheral tissues) is low and activated by injury and/or physico-chemical insult; 2) at the level of its export from cells where it appears to remain cell associated with heparan sulfate related proteoglycans; 3) at the level of these matrix proteoglycans which can serve to sequester the growth factor and keep it away from the target cell; 4) at the level of cell associated proteoglycans which can serve to deliver the growth factor to its signal transducing high affinity receptor; 5) at the level of the high affinity receptor whose expression is exceedingly low in normal quiescent tissues; and finally 6) by the sheer complexity of a ligand receptor complex that can include four genes encoding numerous isoforms of high affinity receptors, seven genes encoding low affinity receptors that are differentially sulfated and glycosylated and at least nine genes encoding different structurally related FGFs.

References

1. Feige J-J, Baird A. Medicine et Sciences 1992;8:805–810.
2. Flaumenhaft R, Rifkin DB. Mol Biol Cell 1992;3:1057–1065.
3. Basilico C, Moscatelli D. Adv Cancer Res 1992;59:115–165.
4. Miyamoto M, Naruo K-I, Seko C, Matsumoto S, Kondo T, Kurokawa T. Mol Cell Biol 1993;13: 4251–4259.
5. Tanaka A, Miyamoto K, Minamino N, Takeda M, Sato B, Matsuo H, Matsumoto K. Proc Natl Acad Sci USA 1992;89:8928–8932.
6. Ornitz DM, Yayon A, Flanagan JG, Svahn CM, Levi E, Leder P. Mol Cell Biol 1992;12:240–247.
7. Rapraeger AC, Krufka A, Olwin BB. Science 1991;252:1705–1708.
8. Klagsbrun M, Baird A. Cell 1991;67:229–231.
9. Patstone G, Pasquale EB, Maher PA. Dev Biol 1993;155:107–123.
10. Lappi DA, Baird A. Progress in Growth Factor Research 1991;2:22–236.
11. Beitz JG, Davol P, Clark JW, Kato J, Medina M, Frackelton AR Jr., Lappi DA, Baird A, Calabresi P. Cancer Res 1992;52:227–230.
12. Lindner V, Lappi DA, Baird A, Majack RA, Reidy MA. Circ Res 1991;68:106–113.
13. Gonzalez AM, Buscaglia M, Ong M, Baird A. J Cell Biol 1990;110:753–765.
14. Baird A, Ling N. Biochem Biophys Res Commun 1987;142:428–435.
15. Vlodavsky I, Folkman J, Sullivan R, Fridman R, Ishai-Michaeli R, Sasse J, Klagsbrun M. Proc Natl Acad Sci USA 1987;84:2292–2296.

16. Folkman J, Klagsbrun M, Sasse J, Wadzinski MG, Ingber D, Vlodavsky I. Amer J Pathol 1988; 130:393–400.
17. Brem S, Tsanaclis AMC, Gately S, Gross JL, Herblin WF. Cancer 1992;70:2673–2680.
18. Hanneken A, De Juan E Jr., Lutty GA, Fox GM, Schiffer S, Hjelmeland LM. Arch Ophthalmol 1991;109:1005–1011.
19. Brunner G, Gabrilove J, Rifkin DB, Wilson EL. J Cell Biol 1991;114:1275–1283.
20. Nguyen M, Watanabe H, Budson AE, Richie JP, Folkman J. JNCI 1993;85:241–242.

Endothelium-Derived Factors and Vascular Functions. T. Masaki, ed.

Coordinative effects of fibroblast growth factor and hypoxanthine on the growth of porcine aortic endothelial cells

A. Ichikawa, Y. Hayashi, S. Hirai, N. Nakanishi, T. Koizumi, Y. Morita and T. Fukui

Department of Physiological Chemistry, Faculty of Pharmaceutical Sciences, Kyoto University, Kyoto 606, Japan

Summary

FGFs had much less effect on growth of PAEC in DMEM with 10% dialyzed FBS. This inhibition was overcome by dialyzable fraction of FBS, and hypoxanthine was purified from it as active compound. Hypoxanthine enhanced the basal and FGF-dependent growth of PAEC. FGFs stimulated PRPP synthesis, and synthesis of purine nucleotides and nucleic acids in association with hypoxanthine via a salvage pathway. Hypoxia induced stimulated production of hypoxanthine, PRPP synthetase/HGPRT activity, FGF synthesis. These data indicate that FGF stimulates the synthesis of PRPP necessary for the salvage synthesis of purine nucleotides in conjunction with purine bases, e.g. hypoxanthine, under hypoxia condition.

Introduction

Fibroblast growth factors (FGFs) stimulate neovascularization in vivo and proliferation and locomotion of endothelial cells in vitro [1]. FGFs are largely produced in proliferating tissues such as follicular cells, placenta, wound healing tissues, and tumor tissues. FGFs are constantly present in normal central nervous tissues such as brain and pituitary [1,2]: however, neovascularization and proliferation of endothelial cells in these tissues are very low [3]. Therefore, it has suspected that the onset of FGF action in these tissues is negatively controlled by some components. Folkman et al. have indicated that basic FGF activity is masked by its binding to heparan sulfate localized in the basement membrane of the bovine cornea [4]. Subsequently, the degradation of basement membrane by heparanase is suspected to induce FGF activation [4]. In contrast to the macromolecular inhibitors, low molecular weight substances such as corticoids, pyrimidine nucleotides, and chemically unidentified component named ESAF [5] have been reported to stimulate endothelial cell proliferation in response to FGF. Recently, we found that hypoxanthine was an essential factor when porcine aortic endothelial cells (PAEC) were cultured in Dulbecco's modified Eagle's medium (DMEM) with 10% dialyzed

172

fetal bovine serum (FBS). Therefore, the aim of the present study was to characterize the coordinative effects of FGF and hypoxanthine on the growth of PAEC. Here we demonstrate that FGF triggers 5-phosphoribosyl 1-pyrophosphate (PRPP) synthesis, resulting in the increased synthesis of purine nucleotides and nucleic acids in association with hypoxanthine via a salvage pathway.

Effect of hypoxanthine on the limited growth of PAEC in the presence of FGFs

Recently, we found that basic FGF or acidic FGF had much less effect on the growth of PAEC in DMEM with 10% dialyzed FBS, compared with that with 10% non-dialyzed FBS [6]. This inhibition was overcome by supplement of dialyzable fraction of FBS, and hypoxanthine was purified from the dialyzable fraction as the active compound which stimulated the basal and FGF-dependent growth rates of dialyzed FBS-treated PAEC [6]. Figure 1 shows the dose-dependent effect of hypoxanthine on the incorporation of [^{3}H]thymidine into DNA of dialyzed FBS-treated PAEC in response to basic FGF. The basal DNA synthetic activity of dialyzed FBS-treated PAEC was stimulated by hypoxanthine, and its activity reached the maximal level at around 5–10 µM of hypoxanthine. In addition, hypoxanthine dose-dependently potentiated the growth-promoting activity of basic FGF, and the maximal activity by 10 ng/ml basic FGF (FGFb) was attained by supplement with around 5–10 µM of hypoxanthine in media, and the plateau level was almost comparable to that potentiated by basic FGF in media with 10% non-dialyzed FBS. These observations suggest that when basic FGF of serum does interact with PAEC, a hypoxanthine is specifically required. However, hypoxanthine

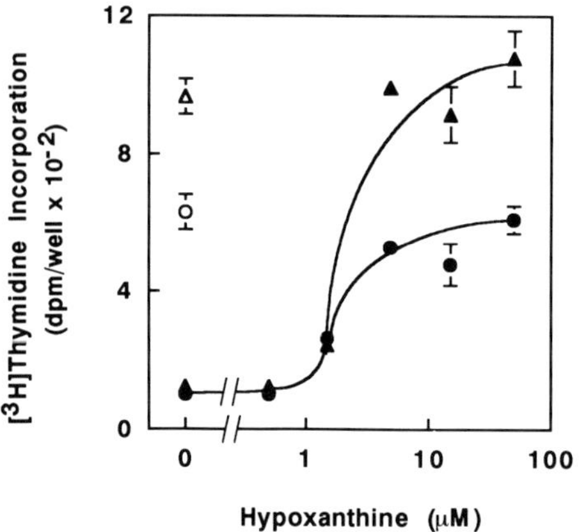

Fig. 1. Effect of hypoxanthine on [^{3}H]thymidine incorporation into DNA of dialyzed FBS-treated PAEC in response to FGFb or FGFa plus heparin. PAEC (1 × 10^4 cells/cm^2) were cultured for 16 h at 37°C in 0.5 ml of media containing 10% FBS (open symbols) or 10% dialyzed FBS (closed symbols) with 10 ng/ml of FGFb (△, ▲) or none (○, ●), and with various doses of hypoxanthine (0.5–50 µM). [^{3}H]Thymidine was added to all wells, and further incubation proceeded for 8 h. Formed [^{3}H]DNA was extracted and counted. Each value is the mean ± SE of four samples.

did not induce more growth-promoting activity in PAEC cultured in 10% non-dialyzed FBS with or without FGFb. The average amount of hypoxanthine in commercially available FBS was about 100 μM [7]. Thus a supplement of 10% FBS into DMEM results in the concentration of hypoxanthine at about 10 μM. We also found that the hypoxanthine concentration in calf serum (CS) or newborn bovine serum (NBS) is about 1/100 of that in FBS, and that the deficiency of growth-promoting activity of CS and NBS on PAEC in DMEM with or without FGFs was due to the low levels of endogenous hypoxanthine [7]. Therefore, in DMEM supplemented with 10 μM hypoxanthine, FGFs enhanced the growth of PAEC in 10% NBS or CS, and the growth rate was almost identical to that in 10% FBS.

In human plasma, hypoxanthine concentration was reported to be around 1–2.5 μM [8]. In addition to growth-promoting activity, we identified angiogenic activity for hypoxanthine, estimated by the chorioallantoic membrane assay. Besides hypoxanthine, the addition of various purine bases and purine nucleosides such as adenine, adenosine, guanine, guanosine, inosine, but not xanthine, xanthosine nor any pyrimidine metabolites, restored the limited growth of PAEC cultured in medium containing 10% dialyzed FBS in the presence or absence of FGFs [9,10].

Purine synthesis salvage pathways in PAEC

Next, we examined the problem what the biological importance of hypoxanthine-induced stimulatory effect on the growth of dialyzed FBS-treated PAEC in response to FGFs might be. In many eucaryotic cell types, hypoxanthine is rapidly salvaged in conjunction with PRPP to yield IMP, which is then converted to adenine or guanine nucleotides through salvage pathways. Consequently, we first speculated that PAEC are specific for the metabolism of purine nucleotides for DNA synthesis at a time of cell proliferation in response to FGFs. The metabolism of [^{14}C]hypoxanthine was compared in PAEC treated with and without FGFs. Treatment of PAEC with FGFs for 24 h enhanced the radioactivity incorporation of [^{14}C]hypoxanthine into both the acid-soluble and -insoluble fractions approximately 2-fold. Upon chromatographic analysis of hypoxanthine metabolites in the acid-soluble nucleotide fraction, it was found that in control cells hypoxanthine was largely metabolized to IMP, adenine nucleotides and uric acid, whereas in FGFs-treated cells it was converted to ATP, ADP, GTP, xanthine, and uric acid. The radioactivity to IMP was lowered in FGFs-stimulated cells. These results indicate that FGFs stimulate a salvage pathway for purine nucleotide synthesis involving hypoxanthine as a substrate. PRPP is used as a substrate for both de novo synthesis and salvage pathways. Subsequently, limitation of PRPP supplementation may regulate purine synthesis. Table 1 shows the effects of FGFs on PRPP content , PRPP synthetase activity, HGPRT activity. The stimulation of PAEC by both FGFb and acidic FGF (FGFa) increased the PRPP content and PRPP synthetase activity 2- to 3-fold and 8- to 10-fold, respectively, but it did not enhance significantly the activity of HGPRT.

Next we examined the role of de novo synthesis of purines in the metabolism of PAEC in response to FGF. Methotrexate, a strong inhibitor of purine synthesis de

174

Table 1. Effects of FGF on the PRPP content, PRPP synthetase activity, and HGPRT activity.

	PRPP content (pmol/10^6 cells)	PRPP synthetase (μU/10^6 cells)	HGPRT activity (mU/10^6 cells)
None	346 ± 22	5.1 ± 0.13	2.96 ± 0.28
FGFa	1160 ± 28[a]	51.4 ± 4.7[a]	3.49 ± 0.35
FGFb	854 ± 17[a]	42.3 ± 2.8[a]	N.D.

PAEC were cultured for 24 h in DMEM supplemented with 10% dialyzed FBS in the presence of 10 ng/ml FGFa plus 10 μg/ml heparin or 10 ng/ml FGFb, and then the PRPP content, HGPRT activity and PRPP synthetase activity were measured. Each value is the mean ± S.E. for three samples. Statistical significance, [a]P < 0.01, N.D., not determined.

novo, exhibited no inhibitory activity towards the growth of PAEC in medium containing 10% dialyzed FBS in the presence or absence of hypoxanthine, and even with 10% non-dialyzed FBS. Furthermore, FGF did not enhance the incorporation of [^{14}C]formate into total purine metabolites in PAEC [9]. These indicate that the de novo pathway may not be preferential for purine synthesis in PAEC. In contrast to purine bases, pyrimidine bases had little effect on the basal or FGF-dependent growth of PAEC. This suggests that pyrimidine synthesis of PAEC depends on pyrimidine synthesis de novo. On the other hand, the addition of FGF to PAEC in dialyzed FBS medium increased PRPP synthetase activity and the PRPP content. Stimulation of PRPP synthetase activity was abolished by simultaneous addition of tyrosine kinase inhibitors, and partially by H-7. Although PMA also increased PRPP synthetase activity, it was specifically inhibited by H-7 but not by tyrosine kinase inhibitors.

Effect of hypoxia on hypoxanthine production, PRPP synthetase activity, HGPRT activity, and FGF synthesis in PAEC

It has been known that irregular nucleotide catabolism and extracellular release of purine metabolites occur in tissues during hypoxia. On these occasions, the endothelium is thought to be one of the candidates that metabolize the released purine metabolites to purine nucleotides and nucleic acids. On the other hand, insufficient vascular supply and the resultant reduction in tissue oxygen tension are known to lead to neovascularization in order to satisfy the needs of the tissue [11]. A number of angiogenic polypeptide growth factors have been demonstrated in vivo. These include FGFb, vascular endothelial growth factor (VEGF), platelet-derived endothelial cell growth factor (PD-ECGF), transforming growth factor-α/epidermal growth factor (TGF-α/EGF), TGFβ, and tumor necrosis factor (TNF-α) [12]. FGFb and/or VEGF, have clearly shown to induce the invasion of endothelial cells grown on three-dimensional collagen or fibrin gels into matrix resulting in the formation of capillary-like tubular structure in vitro model angiogenesis [13,14]. These results demonstrate that angiogenesis-inducing activities of FGFb and/or VEGF can be

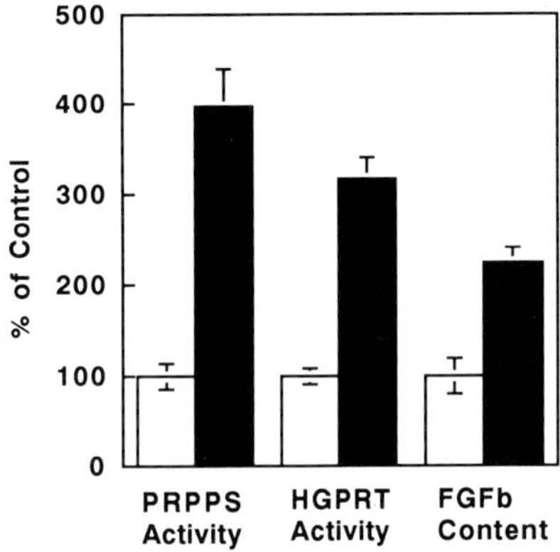

Fig. 2. Effect of hypoxia on PRPP synthetase activity, HGPRT activity and FGF content. PAEC were cultured for 18 h with 95% air/5% CO_2 (control, open columns) or 1% O_2/94% N_2/5% CO_2 (hypoxia, closed columns), and then the enzume activities and FGF content of the cells were assayed. Each value is the mean ± SE of three samples. Control values for PRPP synthetase activity, and HGPRT activity and FGF content were 15.2 ± 2.1 μU/10^6 cells, 1.24 ± 0.11 mU/10^6 cells, and 124 ± 24 pg/10^6 cells, respectively.

mediated via a direct effect on endothelial cells themselves. Hypoxia and adenosine were demonstrated to stimulate the proliferation of bovine aortic endothelial cells in vitro. Although the hypoxia-induced angiogenic factors that mediate marked proliferation of endothelial cells have not been identified, we have examined the effect of hypoxia on FGF-dependent growth of PAEC. PAEC in media with dialyzed FBS were treated with hypoxia (1% O_2/94% N_2/5% CO_2) for 18 h. Compared to control cells (95% air/5% CO_2), hypoxia-treated PAEC consumed much amount of glucose, contained much less amount of ATP, but showed no cytotoxic changes.

On hypoxia treatment, PAEC grown in dialyzed FBS medium enhanced production of hypoxanthine, FGF and FGF mRNA, and de novo synthesis of PRPP synthetase (Fig. 2). FGF-stimulated de novo synthesis of the enzyme was not further potentiated by hypoxia treatment. These data indicate that FGF stimulates the synthesis of PRPP necessary for the salvage synthesis of purine nucleotides in conjunction with purine bases, e.g. hypoxanthine, under hypoxia condition.

Acknowledgements

This work was supported by the grant from the Uehara Memorial Foundation.

References

1. Moscatelli D, Prestan M, Rifkin DB. Proc Natl Acad Sci USA 1986;83:2091–2095.
2. Gospodarowicz D, Mossoglia S, Cheng J, Lui G-M, Baird A, Böhlent P. J Cell Physiol 1985;122: 323–332.
3. Hobson B, Denekamp J. Br J Cancer 1984;49:405–413.
4. Folkman J, Klagsbrun M, Sasse J, Wadzinski M, Ingber D, Vlodavsky I. Am J Pathol 1988;130: 393–400.
5. Odedra R, Weiss JB. Biochem Biophys Res Commun 1987;143:947–953.
6. Hayashi Y, Hirai S, Harayama H, Ichikawa A. Exp Cell Res 1989;185:217–228.
7. Hayashi Y, Hirai S, Ichikawa A. In vitro Cell Dev Biol 1990;26:843–845.
8. Harkness RA. J Chromatogr 1988;429:255–278.
9. Hirai S, Hayashi Y, Koizumi T, Nakanishi N, Fukui T, Ichikawa, A. Biochem Pharmacol 1993;45: 1695–1701.
10. Hayashi Y, Hirai S, Saito T, Harayama H, Ichikawa A. Japan J Pharmacol 1990;53:1–9.
11. Adair TH, Gay WJ, Montani JP. Am J Physiol 1990;259:R393–R404.
12. Klagsbrun M, D'Amore PA. Ann Rev Physiol 1991;53:217–239.
13. Pepper MS, Belin D, Montesano R, Orci L, Vassalli J-D. J Cell Biol 1990;111:743–755.
14. Pepper MS, Ferrara N, Orci L, Montesano R. Biochem Biophys Res Commun 1992;189:824–831.

Endothelium-Derived Factors and Vascular Functions. T. Masaki, ed.

Regulation of the cell cycle of vascular endothelial cells through the protein kinase C pathway

Toshiyuki Sasaguri, Chiya Kosaka, Katsuhiro Zen, Junichi Masuda, Kentaro Shimokado and Jun Ogata

National Cardiovascular Center Research Institute, 5-7-1 Fujishiro-dai, Suita, Osaka 565, Japan

Summary

To elucidate the role of the protein kinase C (PKC) pathway in atherogenesis, we studied PKC-mediated effects on the proliferation of cultured human umbilical vein endothelial cells. Their cell cycle was interrupted by the PKC pathway at two independent points, in the G_1 and the G_2 phases, respectively. The G_1 interruption seemed to be mediated by the inhibition of cyclin A gene expression and pRB phosphorylation, and the G_2 interruption seemed to result from the suppression of Cdc2 kinase activity. We have therefore hypothesized that the PKC pathway contributes to the formation of atherosclerotic lesions by inhibiting endothelial regeneration.

Introduction

According to the response-to-injury hypothesis [1], endothelial cell injury initiates atherosclerotic lesion formation. However, the initial injurious events do not necessarily lead to endothelial denudation, and the earliest manifestations of "injured" endothelial cells can only be alterations in one or more of their physiological properties. Fatty streaks, the early lesions of atherosclerosis, can develop at sites covered with morphologically intact endothelium.

Nevertheless, cell desquamation may happen more frequently in injured than in normal endothelium. When damaged cells are lost and contacts between neighboring endothelial cells are disrupted, cells surrounding the wound migrate into the wounded area to re-establish their cell-cell connections [2]. However, if the wound is too large to be covered by this means, the remaining cells may proliferate. If the recovery from the wound is delayed for any reason, platelets may adhere to the exposed subendothelial tissue and release active substances, such as platelet-derived growth factor and thromboxane A_2, which may promote atherogenesis by stimulating smooth muscle cell migration and proliferation. Leukocytes and plasma components may readily infiltrate the intima and promote lesion formation by providing growth factors, cytokines, and other active agents. In fact, endothelial denudation is observed even during the early phases of atherogenesis [3,4]. Therefore, rapid endothelial regeneration may prevent progression of the lesion. It is important, therefore, to study the way in which endothelial cell proliferation is regulated under both physiological and pathological conditions.

Endothelial cells are exposed to several agents, such as Ca^{2+} mobilizing hormones and growth factors, that stimulate phospholipase-linked receptors. One of the earliest events in their intracellular signaling is the activation of phospholipase C, which hydrolyzes membrane phospholipids to produce 1,2-diacylglycerol (DAG), a physiological activator of protein kinase C (PKC) [5]. Phospholipase D may also be activated to produce phosphatidic acid, which is further hydrolyzed to generate DAG [5]. The PKC pathway can also be activated by other means under pathological conditions. Lysophosphatidylcholine (lysoPC), which accumulates in atherosclerotic vessel walls [6], enhances PKC activity in the presence of DAG in vitro [7]. High concentrations of glucose stimulate PKC by de novo synthesis of DAG [8–10].

The PKC pathway may play several roles in cell proliferation. In some cell systems, PKC activation can promote the G_0/G_1 transition, stimulating the expression of immediate early genes such as c-*fos* and c-*myc* [11–14]. On the other hand, PKC inhibits cell proliferation in several cell species [15–19]. The proliferation of vascular smooth muscle cells is interrupted in the G_1 phase by the PKC pathway [20]. However, PKC-mediated effects on endothelial cell proliferation have yet to be investigated.

The progression of the eukaryotic cell cycle is controlled by several cyclins and cyclin-dependent kinases [21–24]. Cyclin A is expressed from late G_1 to S phase, and has been suggested to play an essential role in G_1/S progression in association with Cdk2, which may subsequently phosphorylate a putative substrate that initiates DNA synthesis [25–29]. Retinoblastoma gene product (pRB) may also be involved in G_1/S regulation. pRB activity is controlled by phosphorylation; the under-phosphorylated (active) form is able to suppress G_1/S transition and the hyper-phosphorylated (inactive) form loses this ability, allowing cells to enter S phase. The underphosphorylated pRB can bind to a transcription factor, E2F, but this ability is lost by phosphorylation [30]. E2F thereby released may stimulate the transcription of several factors essential to DNA replication, such as c-Myc, Cdc2, thymidine kinase, and DNA polymerase-α. G_2/M transition, however, is controlled by B-type cyclins that form a complex with Cdc2 (Cdk1), which phosphorylates several proteins such as nuclear lamins, vimentin, caldesmon, and histone H1, both in vitro and in vivo [23].

Against this background, we studied the role played by the PKC pathway in the regulation of vascular endothelial cell proliferation, with regard to factors controlling the cell cycle, such as cyclins, cyclin-dependent kinases, and pRB [31,32].

Experimental procedures

Cell culture

Human umbilical vein endothelial cells, cultured in Dulbecco's modified Eagle's medium, containing 20% fetal bovine serum (FBS) and 10 ng/ml human recombinant basic fibroblast growth factor (bFGF, Amersham), were used within seven passages. G_0-synchronization was achieved by serum-starvation for 48 h, G_1/S-synchronization was achieved with 1 µg/ml aphidicolin, and M phase-arrest was achieved with 0.5 µg/ml nocodazole or 0.5 µg/ml colcemid. Cell numbers were determined with a Coulter counter.

DNA synthesis assay

Performed as described previously [33].

Flow cytometry

Cells fixed with 70% ethanol were stained with propidium iodide (0.005%), after treatment with RNase (2 mg/ml), or with acridine orange (6 µg per 1 ml of 0.1 M sodium citrate/Na_2HPO_4 [pH 2.6]), after treatment with RNase and 0.1 M KCl/0.1 M hydrochloric acid (pH 1.4). Fluorescence was analyzed with a flow cytometer (Cyto ACE150, Japan Spectroscopic Co.).

Phorbol ester binding and PKC enzyme assays

Performed as described previously [31].

Cyclin-dependent kinase assay

Cells were lysed in 10 mM Tris/HCl (pH 7.4), 1 mM EDTA, 1 mM EGTA, 100 mM NaF, 200 µM Na_3VO_4, 20 µg/ml phenylmethylsulfonyl fluoride, 20 µg/ml leupeptin, and 0.1% Nonidet P-40 (buffer A), and centrifuged at 16,000 × g to remove insoluble pellets. After the addition of 120 mM NaCl, the lysate (1 ml) was incubated with 50 µl of 50% (v/v) protein A Sepharose (Pharmacia) for 30 min at 4°C, followed by centrifugation. The supernatant was incubated with a polyclonal antibody to the carboxyl terminal sequence of Cdc2 kinase (Oncogene Science) for 1 h at 4°C. Then 30 µl of 50% (v/v) protein A Sepharose was added and the incubation was continued for 1 h at 4°C. The precipitate was washed five times with buffer A, after which it was suspended in 20 mM Tris/HCl (pH 7.4) and 10 mM $MgCl_2$. The reaction mixture (75 µl), containing 25 µl of the suspension, 20 mM Tris/HCl (pH 7.4), 10 mM $MgCl_2$, 1 mM dithiothreitol, 100 µg/ml histone H1 (Boehringer), and 50 µM [γ-^{32}P]ATP (Amersham) was incubated for 30 min at 30°C. The reaction was stopped with 30 µl of 300 mM H_3PO_4, and the mixture was then absorbed on Whatman p81 binding paper (9 cm^2). After four washes with 200

180

ml of 75 mM H_3PO_4, radioactivity was determined by liquid scintillation counting. Cdc2-specific activity was calculated by subtracting control values, obtained by reactions in the absence of histone H1, from the activity obtained as outlined above.

Western blotting

Performed basically as described previously [20,31]. When immunoprecipitates were used as samples, they were prepared as described above.

Northern blotting and nuclear run-on assay

Performed as described previously [31,33].

Results

Phorbol 12-myristate, 13-acetate (PMA) inhibited the increase in the endothelial cell population in a dose-dependent manner (Fig. 1). To elucidate the mechanism responsible for this effect, we analyzed the cell cycle of this population. DNA synthesis stimulated with bFGF was potently inhibited by PMA with a dose dependency similar to that obtained for the population increase, the maximal effect (more than 95% inhibition) being obtained at 10 nM. At concentrations above that, the effect was partially reversed. Scatchard analysis for PMA binding showed that [^{3}H]PMA bound to a single affinity site within the concentration range of 1.25–25 nM. 4α-Phorbol, which is unable to activate PKC, had no influence. The PMA effect was completely abrogated by long-term pre-exposure to PMA, which down-regulated PKC activity. 1,2-Dioctanoylglycerol (DiC$_8$), a membrane-permeable DAG, also inhibited DNA synthesis when added repeatedly from the late G$_1$. This effect was abolished, although not completely, in cells pre-exposed to PMA.

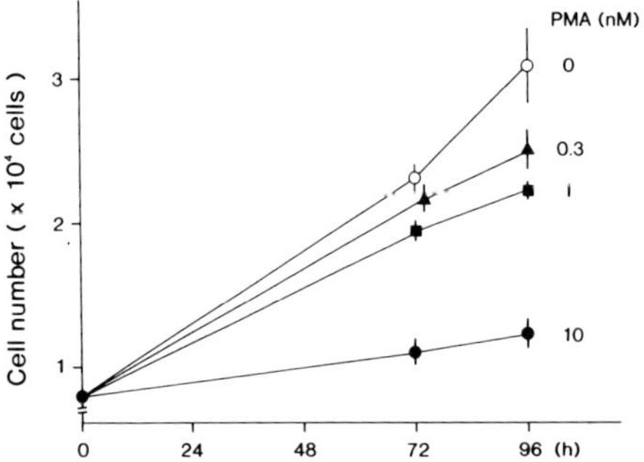

Fig. 1. Effects of PMA on cell population increase. G$_0$-arrested cells were seeded at a density of 8×10^3 per dish (35-mm diameter), and incubated in the presence of 20% FBS, 10 ng/ml bFGF, and various concentrations of PMA. The cells were harvested after 72 or 96 h and counted with a Coulter counter. Each symbol represents the mean ± standard deviation of quadruplicate experiments. Reproduced from [32].

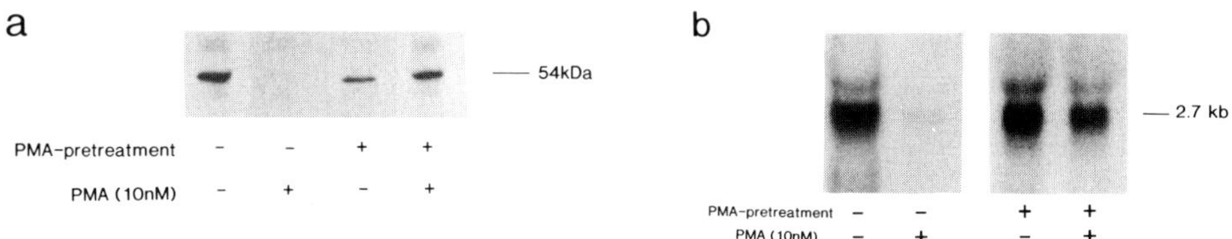

Fig. 2. PKC and cyclin A expression. (a) PMA effect on cyclin A protein expression. PMA (10 nM) was added 8 h after bFGF. The cells were harvested at 20 h and the extracted proteins were analyzed by Western blotting for human cyclin A. Blots were probed with a monoclonal antibody to human cyclin A (C160). (b) PMA effect on cyclin A gene expression. PMA was added 8 h after bFGF, and RNA was extracted at 16 h. Twenty µg of total cellular RNA electrophoresed on 1% agarose gel was hybridized with ^{32}P-labeled human cyclin A cDNA. Reproduced from [32].

The effect of PMA was maximal when applied 8 h after bFGF. This suggests that the check point for PKC is located in the late G_1, since the cells entered S phase about 12 h after the stimulation. This led us to the assumption that PKC influences an event that is essential for the initiation of DNA synthesis and that occurs a few hours before entry into the S phase. The level of cyclin A mRNA was elevated from the late G_1 and reached the maximal value in the early to middle S, followed by a decay. Correspondingly, the level of the protein increased from the late G_1, reaching the maximal value in the middle to late S; it was reduced more slowly than the mRNA. Therefore, the onset of cyclin A gene expression appeared to coincide with the moment of PKC action.

The level of cyclin A protein, determined by Western blotting, was markedly reduced by PMA added 8 h after bFGF, the most effective time for PMA action on DNA synthesis (Fig. 2a). This effect was apparent 4 h after PMA addition and lasted for at least 16 h. The dose dependency was similar to that observed for the proliferation. However, PMA did not inhibit the protein expression in PKC-depleted cells, confirming that the inhibition of cyclin A expression was caused by PKC, and suggesting that a down-regulatable PKC isoform(s) was responsible for this effect. DiC_8 also inhibited the expression. Although DiC_8 was not as potent as PMA, its potency appeared to parallel its effect on proliferation.

To determine whether the decrease in the amount of cyclin A protein was due to suppression of its gene expression, we performed Northern blotting. The level of cyclin A mRNA was potently reduced by PMA (Fig. 2b). The dose dependency was, again, similar to that shown for the PMA effect on DNA synthesis. However, PMA failed to reduce the mRNA levels in PKC-depleted cells. DiC_8 also reduced the mRNA levels, with a similar dose-dependency to that shown for the protein levels. To examine whether the decrease in cyclin A mRNA was caused by inhibition of transcription, we performed a nuclear run-on assay. PMA inhibited the nascent transcription of the mRNA, indicating that cyclin A transcription was negatively regulated through the PKC pathway.

Figure 3 demonstrates the effect of PMA on the phosphorylation of pRB. Hyperphosphorylated pRB moved more slowly in SDS-polyacrylamide gel than the underphosphorylated form. Underphosphorylated pRB was predominant in G_0-

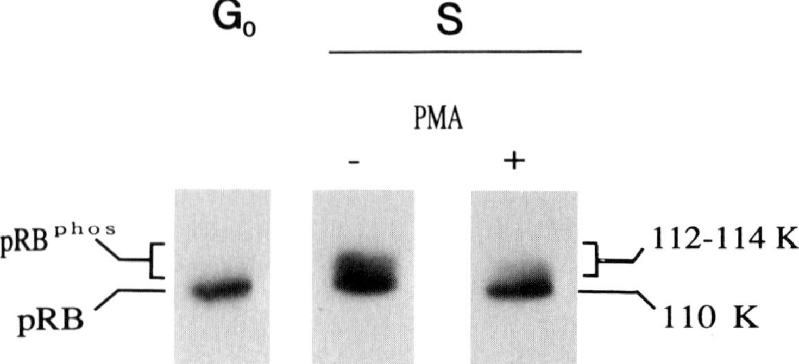

Fig. 3. Effects of PMA on pRB phosphorylation. PMA (10 nM) was added 8 h after bFGF, and proteins were extracted at 20 h. The proteins immunoprecipitated with anti-pRB monoclonal antibody (C36) were immunoblotted with another anti-pRB monoclonal antibody (PMG3-245). Reproduced from [32].

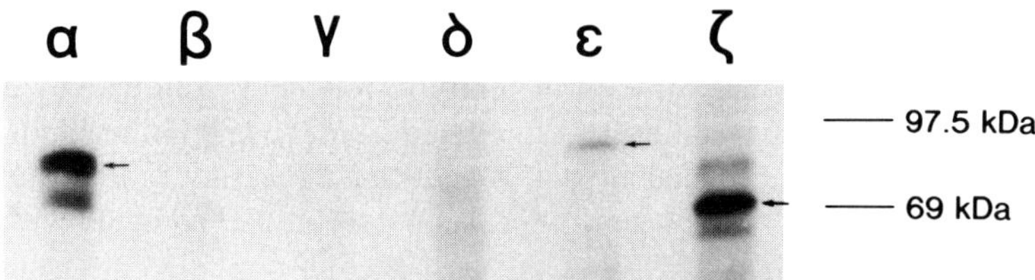

Fig. 4. PKC isoforms expressed in endothelial cells. Soluble proteins extracted from exponentially growing cells were immunoblotted with antibodies specific to PKC isoforms; monoclonal antibodies MC-3a, MC-2a, and MC-1a for PKC-α, -β, and -γ, respectively, and polyclonal antibodies (GIBCO/BRL) for the others. Reproduced from [32].

arrested cells, whereas the hyperphosphorylated form increased after the cells entered S phase. PMA added in the late G_1 markedly inhibited this phosphorylation.

We then examined whether the G_2/M transition was also regulated by the PKC pathway. Cells arrested at the G_1/S boundary by aphidicolin, released by washing out the inhibitor, were incubated in the growth medium. DNA synthesis was completed by 12 h after the release, and the cells entered M phase at 24–27 h. PMA added by 21 h strongly inhibited the increase in the cell population. However, PMA did not interrupt completion of the M phase in cells released from the nocodazole-induced M phase-arrest. Flow cytometry revealed that cells in the G_2 or M phases accumulated in the presence of PMA, which finding suggests that the PKC pathway does not interrupt progression of the S phase. Furthermore, cells did not enter M phase in the presence of both PMA and colcemid, an M phase inhibitor, whereas, in the presence of colcemid alone, cells in the M phase accumulated markedly. This indicates that the check point for PMA is located in the G_2, and not in the M phase.

To elucidate the mechanism responsible for the G_2 inhibition, we investigated the effect of PMA on Cdc2 kinase activity. Cdc2 was activated around the G_2/M border.

However, in the presence of PMA, this activation was strongly suppressed, in a concentration-dependent manner.

To determine the PKC isoform(s) involved in the cell cycle inhibition, we conducted Western blotting, using antibodies specific to the isoforms. From the soluble proteins of the endothelial cells, PKC-α, -ε, and -ζ were detected (Fig. 4).

Discussion

We found that the cell cycle of cultured human vascular endothelial cells was negatively regulated by the PKC pathway at two independent points, in the G_1 and the G_2 phases, respectively (Fig. 5). That this inhibition did not abrogate the net effect of PKC activation on cell proliferation was shown by the finding that the increase in cell population was also inhibited by the PKC activators. However, it appears that PKC-mediated growth inhibition is not a phenomenon limited to endothelial cells, and that it may be a more universal mechanism, since PKC has been reported to inhibit proliferation in other cell systems as well [15–19]. As far as our studies are concerned, G_1/S transition was always inhibited, regardless of the cell species used in the experiments, which species included human, porcine, bovine, and rat arterial smooth muscle cells, human umbilical vein and bovine aortic endothelial cells, and the human lung fibroblast line, IMR-90. Although PKC does

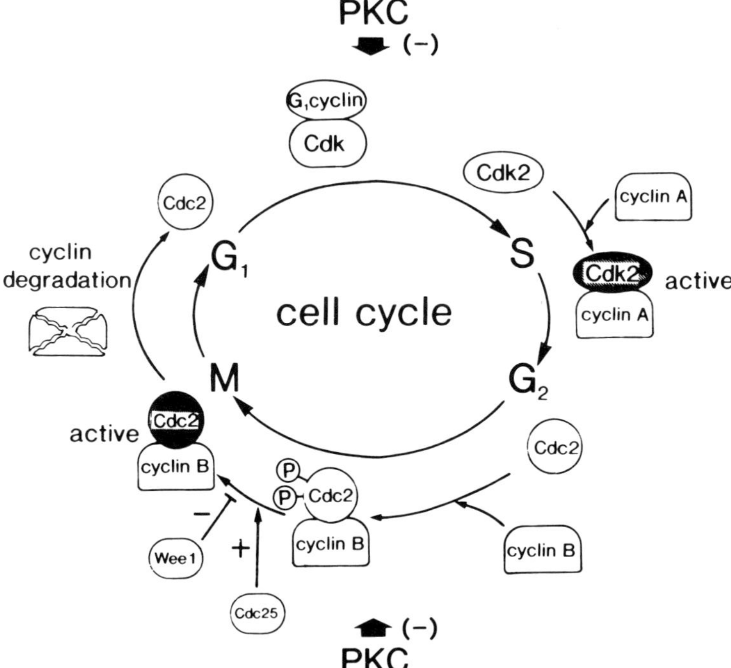

Fig. 5. Cell cycle regulation by the PKC pathway.

186

proportion to excess extracellular glucose. An excess of intracellular glucose may lead to activation of the polyol pathway, generating fructose via sorbitol, elevating the NADH/NAD$^+$ ratio, which process may result in an increase in the de novo synthesis of DAG via glycerol 3-phosphate [9]. The PKC pathway thereby activated could be involved in the glucose-mediated impairment of endothelial cell proliferation, and therefore could be a cause of diabetic angiopathies.

In vitro at least, the PKC pathway of vascular endothelial cells can be activated on several occasions, as outlined above. If it is also activated in vivo, endothelial regeneration after injury may be interrupted, resulting in the retardation of intimal repair and therefore in accelerated atherosclerotic lesion formation. We conclude with our hypothesis that the PKC pathway contributes to atherogenesis by inhibiting endothelial regeneration (Fig. 6). This would modify the response-to-injury hypothesis.

Acknowledgement

This study was supported by grants from the Science and Technology Agency, the Ministry of Health and Welfare, Japan, the Uehara Memorial Foundation, and the Yamanouchi Foundation for Research on Metabolic Disorders.

References

1. Ross R. Nature (London) 1993;362:801–809.
2. Reidy MA. Lab Invest 1985;53:513–520.
3. Davies MJ, Woolf N, Rowles PM, Pepper J. Br Heart J 1988;60:459–464.
4. Masuda J, Ross R. Arteriosclerosis 1990;10:164–177.
5. Nishizuka Y. Science 1992;258:607–614.
6. Portman OW, Alexander M. J Lipid Res 1969;10:158–165.
7. Asaoka Y, Oka M, Yoshida K, Sasaki Y, Nishizuka Y. Proc Natl Acad Sci USA 1992;89:6447–6451.
8. Craven PA, DeRubertis FR. J Clin Invest 1989;83:1667–1675.
9. Lee T-S, Saltsman KA, Ohashi H, King GL. Proc Natl Acad Sci USA 1989;86:5141–5145.
10. Ayo SH, Radnik R, Garoni JA, Troyer DA, Kreisberg JI. Am J Physiol 1991;261:F571–F577.
11. Dicker P, Rozengurt E. Nature (London) 1980;287:607–612.
12. Rozengurt E, Rodriguez-Pena A, Coombs M, Sinnett–Smith J. Proc Natl Acad Sci USA 1984;81:5748–5752.
13. Greenberg ME, Ziff EB. Nature (London) 1984;311:433–438.
14. Reed JC, Nowell PC, Hoover RG. Proc Natl Acad Sci USA 1985;82:4221–4224.
15. Rovera G, Santoli D, Damsky C. Proc Natl Acad Sci USA 1979;76:2779–2783.
16. McKay I, Collins M, Taylor–Papadimitriou J, Rozengurt E. Exp Cell Res 1983;145:245–254.
17. Huang C-L, Ives HE. Nature (London) 1987;329:849–850.
18. Kanakura Y, Druker B, DiCarlo J, Cannistra SA, Griffin JD. J Biol Chem 1991;266:490–495.
19. Mackanos EA, Pettit GR, Ramsdell JS. J Biol Chem 1991;266:11205–11212.
20. Sasaguri T, Kosaka C, Hirata M, Masuda J, Shimokado K, Fujishima M, Ogata J. Exp Cell Res 1993;208:311–320.
21. Wang J, Chenivesse X, Henglein B, Bréchot C. Nature (London) 1990;343:555–557.
22. Hunter T, Pines J. Cell 1991;66:1071–1074.

23. Norbury C, Nurse P. Annu Rev Biochem 1992;61:441-470.
24. Sherr CJ. Cell 1993;73:1059–1065.
25. Tsai L-H, Harlow E, Meyerson M. Nature (London) 1991;353:174–177.
26. Girard F, Strausfeld U, Fernandez A, Lamb NJC. Cell 1991;67:1169–1179.
27. Pagano M, Pepperkok R, Verde F, Ansorge W, Draetta G. EMBO J 1992;11:961–971.
28. Rosenblatt J, Gu Y, Morgan DO. Proc Natl Acad Sci USA 1992;89:2824–2828.
29. Elledge SJ, Richman R, Hall FL, Williams RT, Lodgson N, Harper JW. Proc Natl Acad Sci USA 1992;89:2907–2911.
30. Nevins JR. Science 1992;258:424–429.
31. Kosaka C, Sasaguri T, Masuda J, Zen K, Shimokado K, Yokota T, Ogata J. Biochem Biophys Res Commun 1993;193:991–998.
32. Kosaka C, Sasaguri T, Zen K, Masuda J, Shimokado K, Ogata J. Jpn Circ J 1993;57 Suppl. IV: 1207–1210.
33. Sasaguri T, Masuda J, Shimokado K, Yokota T, Kosaka C, Fujishima M, Ogata J. Exp Cell Res 1992;200:351–357.
34. Akiyama T, Ohuchi T, Sumida S, Matsumoto K, Toyoshima K. Proc Natl Acad Sci USA 1992;89:7900–7904.
35. Schwarz JK, Devoto SH, Smith EJ, Chellappan SP, Jakoi L, Nevins JR. EMBO J 1993;12:1013–1020.
36. Lin CR, Chen WS, Lazar CS, Carpenter CD, Gill GN, Evans RM, Rosenfeld MG. Cell 1986;44:839–848.
37. Leach KL, Ruff VA, Jarpe MB, Adams LD, Fabbro D, Raben DM. J Biol Chem 1992;267:21816–21822.
38. James G, Olson E. J Cell Biol 1992;116:863–874.
39. Buchner K, Otto H, Hilbert R, Lindschau C, Haller H, Hucho F. Biochem J 1992;286:369–375.
40. Pfeffer LM, Eisenkraft BL, Reich NC, Improta T, Baxter G, Daniel-Issakani S, Strulovici B. Proc Natl Acad Sci USA 1991;88:7988–7992.
41. Resnitzky D, Tiefenbrun N, Berissi H, Kimchi A. Proc Natl Acad Sci USA 1992;89:402–406.
42. Cocks BG, Vairo G, Bodrug SE, Hamilton JA. J Biol Chem 1992;267:12307–12310.
43. Parthasarathy S, Steinbrecher UP, Barnett J, Witztum JL, Steinberg D. Proc Natl Acad Sci USA 1985;82:3000–3004.
44. Quinn MT, Parthasarathy S, Steinberg D. Proc Natl Acad Sci USA 1988;85:2805–2809.
45. Kume N, Cybulsky MI, Gimbrone MA Jr. J Clin Invest 1992;90:1138–1144.
46. Kugiyama K, Kerns SA, Morriset JD, Roberts R, Henry PD. Nature (London) 1990;344:160–162.
47. Stout RW. Diabetologia 1982;23:436-439.
48. Lorenzi M, Cagliero E, Toledo S. Diabetes 1985;34:621–627.

Receptors and signal transduction in vascular systems

Signal transduction and ligand-binding domains of the tachykinin receptors

S. Nakanishi[1], Y. Nakajima[1], Y. Yokota[1], U. Gether[2] and T.W. Schwartz[2]

[1]*Institute for Immunology, Kyoto University Faculty of Medicine, Kyoto 606, Japan; and* [2]*Laboratory of Molecular Endocrinology, University Department of Clinical Biochemistry, Copenhagen 2100, Denmark*

Summary

The mammalian tachykinin receptors consist of the substance P, substance K and neuromedin K receptors and belong to the family of G protein-coupled receptors. By expressing the individual receptor cDNAs in heterologous CHO cells, we demonstrate that the tachykinin receptors have the potential to couple directly to both phospholipase C and adenylate cyclase and to stimulate phosphatidylinositol hydrolysis and cAMP formation. We also disclose that the transmembrane segment V–VII region contributes to interaction with the common tachykinin sequence, while its preceding amino-terminal portion is responsible for recognition of the divergent tachykinin sequences. In contrast, non-peptide antagonists act through the restricted regions around transmembrane segment VI, indicating the different modes of bindings of natural peptides and non-peptide antagonists.

Introduction

The family of tachykinin peptides consists of three distinct peptides, substance P (SP), substance K (SK) and neuromedin K (NK), and possesses three different types of receptors (SPR, SKR and NKR), each specific for the respective peptide [1,2]. The members of this peptide family are widely distributed both in the central nervous system and in peripheral tissues and evoke a variety of biological activities including smooth muscle contraction, vasodilation, hypotension and excitation of neurotransmission [2]. The involvement of tachykinins in the pathology of certain human diseases, especially inflammatory diseases such as arthritis, has also been suggested by studies of animal models and humans [2]. For the past few years, we have been working on the molecular characterization of the tachykinin receptors. This article summarizes our studies dealing with the molecular structures, signal transduction mechanisms and ligand-binding domains of the three rat tachykinin receptors.

Molecular cloning and structures of tachykinin receptors

Because nothing was known about the molecular nature of the tachykinin receptors, we first attempted to identify the molecular entity of the tachykinin receptor by

one-order higher concentrations of peptides than the stimulation of phosphatidylinositol (PI) hydrolysis [7].

The direct linkage of the tachykinin receptors to both phospholipase C and
adenylate cyclase was examined by investigating the effects of tachykinin
application on these enzyme activities in membrane preparations isolated from
receptor-expressing cells [7] (Fig. 3). Tachykinin, added together with GTP,
activated the activities of both phospholipase C and adenylate cyclase in the
membrane preparations [7]. The rank orders of potencies of the tachykinin peptides
for the activation of phospholipase C and adenylate cyclase were in good
accordance with the selectivity of peptide binding of the respective receptors. In
addition, about a one-order of higher concentrations of peptides was required for the
activation of adenylate cyclase than that for phospholipase C in all three receptors.
Our investigations thus demonstrate that the tachykinin receptors have the potential
to couple directly to both phospholipase C and adenylate cyclase and to stimulate
PI hydrolysis and cyclic AMP formation (Fig. 4).

Ligand-binding domains

The structural basis for the peptide-receptor interaction is important for understand-

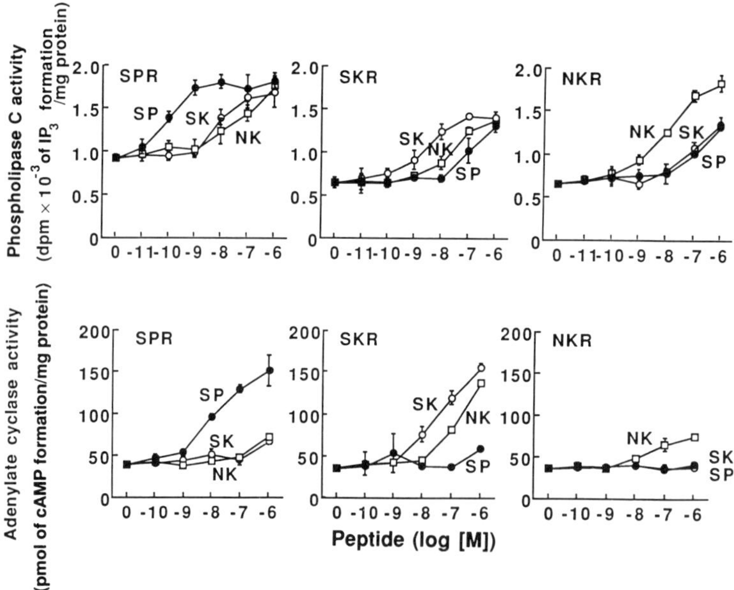

Fig. 3. Dose-response curves of the three tachykinin peptides for the activation of phospholipase C
(upper) and adenylate cyclase (lower) activities in membrane preparations of CHO cells expressing the
individual cloned tachykinin receptors. Indicated concentrations of a tachykinin were incubated with
membrane preparations of receptor-expressing cells in the presence of 100 μM GTP for 10 min. IP₃ and
cyclic AMP formations were then determined. Data are taken from [7].

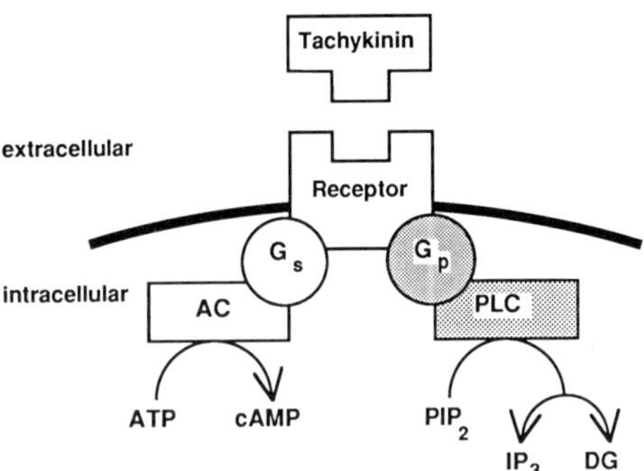

Fig. 4. Signal transduction coupling to the PI hydrolysis and cyclic AMP cascades of the tachykinin receptors expressed in CHO cells. Gs and Gp, G proteins; AC, adenylate cyclase; PLC, phospholipase C; PIP_2, phosphatidylinositol 4,5-biphosphates; DG, diacylglycerol.

ing the function of the tachykinin receptors. We investigated the ligand-binding domains for both natural peptides and synthetic non-peptide antagonists. To map the peptide-binding domains, we constructed a series of chimeric receptors [8] by exchanging systematically seven transmembrane segments between the rat SPR and SKR [8]. We then determined the subtype specificity of the chimeric receptors by radioligand binding and PI hydrolysis measurements after transfection into COS cells [8]. In both analyses, the presence of the middle portion covering transmembrane segment II to the second extracellular loop of SPR conferred a high affinity binding to substance P (Fig. 5). In addition, the exchange of the extracellular amino-terminal domain affected the affinity of SPR. However, replacement of the carboxyl-terminal region up to transmembrane segment V of SPR with the corresponding region of SKR retained a subtype specificity of SPR. This analysis led to the conclusion that the selectivity of the tachykinin receptors is mainly determined by the region extending from transmembrane segment II to the second extracellular loop together with a minor contribution of the extracellular amino-terminal domain, while the region covering transmembrane segments V–VII is responsible for recognition of the common tachykinin sequence [8].

Our previous study of receptor binding with various tachykinin peptides and their related fragments indicated that a fundamental binding and a high-affinity binding of the tachykinin peptides originate from the common carboxyl-terminal sequence and the divergent amino-terminal sequences of the tachykinin peptides, respectively [9]. This structure-activity relationship of the tachykinin peptides coincides well with the concept of the "message-address" sequence of the peptide [9]. The common tachykinin sequence corresponds to a message sequence, while the divergent amino-terminal sequence represents an address sequence. In relation to the "message-address" concept, it can be postulated that the region covering transmembrane segments V–VII contributes to the interaction with the message sequence, while its preceding amino-terminal portion is involved in recognition of the address sequence of the tachykinin peptides (Fig. 6). This structural map may be much too simplified,

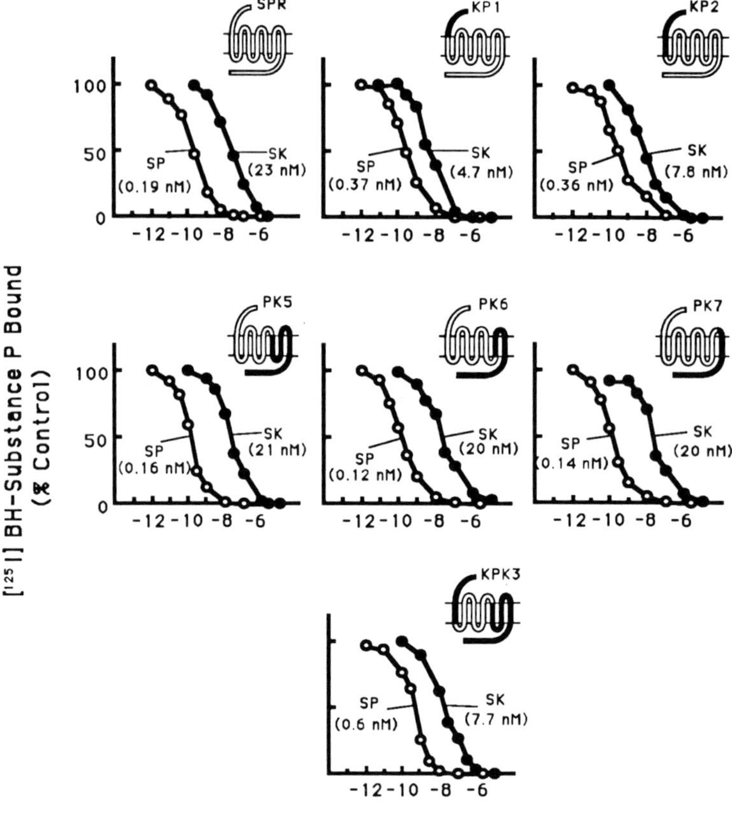

Fig. 5. Displacements of [^{125}I]Bolton-Hunter substance P binding to SPR and chimeric receptors by substance P (SP) and substance K (SK). IC$_{50}$ values of various receptors are indicated in parentheses; Note that the IC$_{50}$ values of SKR for substance P and substance K were determined to be 110 nM and 0.33 nM, respectively. In insets, the sequence derived from SPR is indicated by an open line, while that derived from SKR is displayed by a solid line. Data are taken from [8].

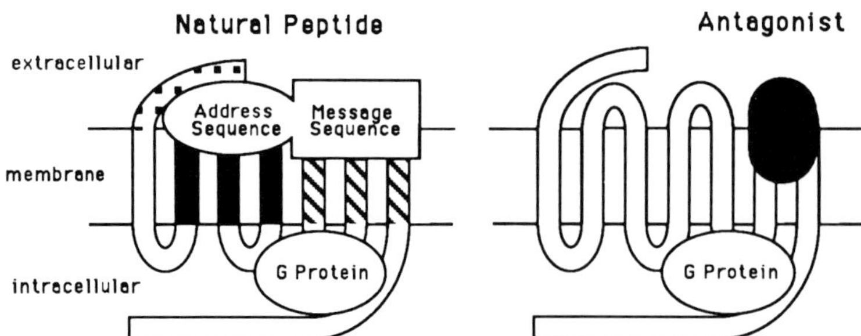

Fig. 6. A model of the ligand-binding domains of natural peptides and non-peptide antagonists of the tachykinin receptors. For non-peptide antagonists, the binding site of the SKR-selective SR-48,968 is indicated.

but this model would be useful for understanding the function of the tachykinin receptors.

Recently, highly potent and selective non-peptide antagonists have been developed for both SPR and SKR. CP-96,345 and SR-48,968 are representative non-peptide antagonists that act specifically on SPR and SKR, respectively, and both pharmacologically serve as competitive antagonists. The mechanism of action of these compounds, however, was obscure because they do not chemically resemble the natural peptides. Collaboration work between the group of Schwartz and our group was initiated to investigate the mechanism of the binding of two chemically very different and highly specific antagonists, CP-96,345 and SR-48,968 [10,11]. We studied the binding domains of the tachykinin receptors to these non-peptide antagonists by utilizing the above and additionally constructed chimeric receptors between the SPR and SKR or between the SPR and NKR. The results indicated that non-conserved residues, in two regions at the top of transmembrane segment V and in one region at the top of transmembrane segment VI, are essential for the specific binding of CP-96,345 to the SPR [10]. The binding site for SR-48,966 was similarly localized at the region corresponding to transmembrane segment VI, the amino-terminal half of transmembrane segment VII and the connecting third extracellular loop of the SKR [11]. Our data thus demonstrate that the two non-peptide antagonists act through the limited regions around transmembrane segment VI on their respective target receptors despite the fact that these regions are not important for determining the ligand-binding specificity of the natural peptides.

Concluding remarks

This article describes our molecular analysis of the structures, signal transduction mechanisms and ligand-binding domains of the tachykinin receptors. The tachykinin receptors belong to the family of G protein-coupled receptors and exert a variety of their functions through the activation of the characteristic second messenger system. The mRNAs for the three tachykinin receptors are differently distributed in peripheral tissues and the central nervous system [12]. The expression of the tachykinin receptor genes [13] is also regulated in response to external stimuli [14]. The tachykinin receptors thus exhibit their functional diversity by employing various cellular mechanisms characteristic of eukaryotic cells.The structural basis of ligand binding for both natural peptides and non-peptide antagonists has also be reported. Our finding that the non-peptide antagonists act through non-conserved residues at limited regions of the receptors implies that the binding pockets for natural peptides and non-peptide antagonists overlap spatially but do not utilize the same set of amino acid residues for their binding to the tachykinin receptors.

Acknowledgements

We thank our colleagues in Institute for Immunology, Kyoto University Faculty of Medicine and Laboratory of Molecular Endocrinology, University Department of Clinical Biochemistry, the Uehara Memorial Foundation, and the Ministry of Education, Science and Culture of Japan.

References

1. Nakanishi S, Ohkubo H, Kakizuka A, Yokota Y, Shigemoto R, Sasai Y, Takumi T. Rec Prog Horm Res 1990;46:59–84.
2. Nakanishi S. Annu Rev Neurosci 1991;14:123–136.
3. Masu Y, Nakayama K, Tamaki H, Harada Y, Kuno M, Nakanishi S. Nature 1987;329:836–838.
4. Sasai Y, Nakanishi S. Biochem Biophys Res Commun 1989;165:695–702.
5. Yokota Y, Sasai Y, Tanaka K, Fujiwara T, Tsuchida K, Shigemoto R, Kakizuka A, Ohkubo H, Nakanishi S. J Biol Chem 1989;264:17649–17652.
6. Shigemoto R, Yokota Y, Tsuchida K, Nakanishi S. J Biol Chem 1990;265:623–628.
7. Nakajima Y, Tsuchida K, Negishi M, Ito S, Nakanishi S. J Biol Chem 1992;267:2437–2442.
8. Yokota Y, Akazawa C, Ohkubo H, Nakanishi S. EMBO J 1992;11:3585–3591.
9. Ingi T, Kitajima Y, Minamitake Y, Nakanishi S. J Pharmacol Exp Ther 1991;259:968–975.
10. Gether U, Yokota Y, Emonds-Alt X, Brelière J-C, Lowe III JA, Snider RM, Nakanishi S, Schwartz TW. Proc Natl Acad Sci USA 1993;90:6194–6198.
11. Gether U, Johansen TE, Snider RM, Lowe III JA, Nakanishi S, Schwartz TW. Nature 1993;362:345–348.
12. Tsuchida K, Shigemoto R, Yokota Y, Nakanishi S. Eur J. Biochem 1990;193:751–757.
13. Takahashi K, Tanaka A, Hara M, Nakanishi S. Eur J Biochem 1992;204:1025–1033.
14. Ihara H, Nakanishi S. J Biol Chem 1990;265:22441–22445.

Endothelium-Derived Factors and Vascular Functions. T. Masaki, ed.

Cloning, expression and regulation of angiotensin II receptors

Tadashi Inagami, Naoharu Iwai, Katsutoshi Sasaki, Yoshiaki Yamano, Smriti Bardhan, Shigeyuki Chaki, Deng-Fu Guo, Hiroaki Furuta, Kenji Ohyama, Yoshikazu Kambayashi, Kyoko Takahashi and Toshihiro Ichiki

Department of Biochemistry, Vanderbilt University School of Medicine, Nashville, TN 37232-0146, U.S.A.

Summary

Angiotensin II isoform 1 (AT_1) receptor cDNAs were cloned by expression cloning from bovine adrenal rat kidney and rat vascular smooth muscles. Human AT_1 receptor was also cloned. A seven transmembrane structure emerged. A single type receptor seems to interact with more than one type G-proteins. AT_1 consists of subtypes AT_{1A} and AT_{1B}. The regulation of the receptors occur at many stages. The isoform, AT_2, was also expression cloned from rat pheochromocytoma cells. Although its ligand binding is not affected by GTP analogs, it is a seven transmembrane domain receptor. It mediates the inhibition of phosphotyrosine phosphatase by angiotensin II and AT_2 specific CGP42112A. The inhibition was abolished by pertussis toxin. Thus, AT_2 belongs to a new class of angiotensin receptors with unique signalling and regulatory mechanisms.

Introduction

The diversity of ANG II mediated actions encompasses numerous tissues and many different modes of action with different durations, and the existence of multiple receptors has been postulated to explain diverse cellular signalling mechanisms [1]. Indeed invention of subtype-specific inhibitors revealed the presence of three major groups of angiotensin II receptors, AT_1, AT_2 and non AT_1-non AT_2 (AT_3) [2-6]. Yet, studies with subtype specific inhibitors indicate that practically all of the known responses elicited by angiotensin, either short- or long-term, are mediated by the angiotensin II type 1 (AT_1) receptor subtype [7]. Therapeutic studies using various inhibitors of the renin-angiotensin system have indicated that this system is important in a large subpopulation of patients with essential hypertension. Again, AT_1-specific receptor antagonists are effective in correcting blood pressure in these hypertensive populations. The molecular basis of the receptor diversity has been very difficult to determine. Attempts to purify solubilized receptors have failed because angiotensin II receptors are very unstable. The problem has been compounded by the possibility that the *mas* oncogene product may be an angiotensin II receptor [8] and by the finding of stable cytosolic angiotensin II binding proteins [9,10].

200

```
                                                                     I
Rat AT1A     MALNSSAEDG  IKRIQDDCPK  AGRHSYIFVM  IPTLYSIIFV  VGIFGNSLVV    50
Rat AT1B     -T----T---  ----------  ----------  ----------  ----------
Mouse AT1A   ------T---  ---------R  ----------  ----------  ----------
Mouse AT1B   -I----I---  ----------  ----N-----  ----------  ----------
Rabbit AT1   -M----T---  ----------  ----N-----  ----------  --------A-
Bovine AT1   -I----T---  ----------  ----N---I-  ----------  ----------
Human AT1    -I----T---  ----------  ----------  ----------  ----------

                                     II
Rat AT1A     IVIYFYMKLK  TVASVFLLNL  ALADLCFLLT  LPLWAVYTAM  EYRWPFGNHL   100
Rat AT1B     ----------  ----------  ----------  ----------  ----------
Mouse AT1A   ----------  ----------  ----------  ----------  ----------
Mouse AT1B   ----------  ----------  ----------  ----------  --Q-------
Rabbit AT1   ----------  ----------  ----------  ----------  --------Y-
Bovine AT1   ----------  ----------  ----------  ----------  --------Y-
Human AT1    ----------  ----------  ----------  ----------  --------Y-

                         III
Rat AT1A     CKIASASVTF  NLYASVFLLT  CLSIDRYLAI  VHPMKSRLRR  TMLVAKVTCI   150
Rat AT1B     --------S-  ----------  ----------  ----------  ----------
Mouse AT1A   --------S-  ----------  ----------  ----------  ----------
Mouse AT1B   --------S-  ----------  ----------  ----------  ----------
Rabbit AT1   --------S-  ----------  ----------  ----------  ----------
Bovine AT1   --------S-  ----------  ----------  ----------  ----------
Human AT1    --------S-  ----------  ----------  ----------  ----------

                  IV                        ▼           ▼
Rat AT1A     IIWLMAGLAS  LPAVIHRNVY  FIENTNITVC  AFHYESRNST  LPIGLGLTKN   200
Rat AT1B     ----------  -----Y----  ----------  ------Q---  ----------
Mouse AT1A   ----------  ----------  ----------  ----------  ----------
Mouse AT1B   ----------  ----------  ----------  ------Q---  ----------
Rabbit AT1   ----L-----  ---I-----F  ----------  ------Q---  ----------
Bovine AT1   ----L-----  --TI-----F  ----------  ------Q---  --V-------
Human AT1    ----L-----  ---I-----F  ----------  ------Q---  ----------

                  V
Rat AT1A     ILGFLFPFLI  ILTSYTLIWK  ALKKAYEIQK  NKPRNDDIFR  IIMAIVLFFF   250
Rat AT1B     ----V-----  ----------  -------K---  -T--------  ----------
Mouse AT1A   ----------  ----------  ----------  ----------  ----------
Mouse AT1B   ----V---V-  ----------  -------K---  -T--------  ----------
Rabbit AT1   ----------  ----------  ----------  ---------K  ----------
Bovine AT1   ----------  ----------  T---------  ----K----K  --L-------
Human AT1    ----------  ----------  ----------  ---------K  ----------

                  VI                                        VII
Rat AT1A     FSWVPHQIFT  FLDVLIQLGV  IHDCKISDIV  DTAMPITICI  AYFNNCLNPL   300
Rat AT1B     ----------  ---------I  -R--E-A---  ----------  ----------
Mouse AT1A   ----------  ----------  ------A---  ----------  ----------
Mouse AT1B   ---------S  ----------  ----E-A-V-  ----------  ----------
Rabbit AT1   ----------  ----------  ----R-A---  ----------  ----------
Bovine AT1   ----------  -M-------L  -R----E---  ---------L  ----------
Human AT1    ---I------  ---------I  -R--R-A---  ----------  ----------

                  
Rat AT1A     FYGFLGKKFK  KYFLQLLKYI  PPKAKSHSSL  STKMSTLSYR  PSDNMSSSAK   350
Rat AT1B     ----------  ----------  --T----AG-  ----------  ----------
Mouse AT1A   ----------  ----------  ----------  ----------  -------A--
Mouse AT1B   ----------  R---------  ----------  ----------  ---------R
Rabbit AT1   F---------  ----------  --------N-  ----------  ----V---S-
Bovine AT1   ----------  ----------  --------N-  ----------  --E-GN--T-
Human AT1    ----------  R---------  --------N-  ----------  ----V---T-

Rat AT1A     KPASCFEVE*                                                   359
Rat AT1B     -S--F-----
Mouse AT1A   -----S----
Mouse AT1B   -S-Y------
Rabbit AT1   --VP------
Bovine AT1   ---P-I----
Human AT1    ---P------
```

Fig. 1. Amino acid sequences of AT$_1$ receptors deduced from the base sequences of cloned cDNA.

Cloning and structure of angiotensin II receptors

Recent development of the expression cloning technique using mammalian cells [11] has now made it possible to address longstanding issues concerning the structure of these unstable and complex receptors. Thus, AT_1 (type 1 angiotensin II receptor) was expression-cloned from cultured bovine adrenocortical cells [12] and rat vascular smooth muscle cells [13], which was rapidly followed by hybridization cloning of rat AT_1 cDNA from the kidney of spontaneously hypertensive rats [14], human hepatic cDNA, human genomic DNA and mouse genomic DNA (Fig. 1). In rodents, rat pituitary and adrenal tissue contained a variant of AT_1, which are now designated as AT_{1B} to distinguish it from the vascular, renal and hepatic receptor AT_{1A} [15,16]. Both AT_{1A} and AT_{1B} were found to have typical structural features of a seven-transmembrane domain receptor. Although these receptors are closely related in base or amino acid sequence (approximately 95% identical in amino acid sequence). However, there seems to be important structural differences in the third cytosolic loop, which plays an important role in interacting with G-proteins.

Type 2 angiotensin II receptor, AT_2, had been considered to be unrelated to AT_1 because of its lack of response to GTP and its stable analog GTPγS [17-20]. This indicated the possible absence of coupling to a G-protein. Its mRNA did not show cross-reaction with cDNA for AT_1. Again, expression-cloning was called for, but the rapid mutation of AT_2 cDNA in various expression vectors during manipulation presented considerable difficulty to the cloning work. Using conditions to minimize the mutation and stringent size separation of cDNA by K-acetate gradient

```
AT1               MALN   SSAEDGIKRI  QDDCPKAGRH  SYIFVMIPTL
AT2   MKDNFSFAAT   SRNITSSLPF  DNLNATGTNE  SAFNCSHKPA  DKHLEAIPVL   50

AT1   YSIIFVVGIF   GNSLVVIVIY  FYMKLKTVAS  VFLLMLALAD  LCFLLTLPLW
AT2   YYMIFVIGFA   VNIVYVSLFC  CQKGPKKVSS  IYIFNLAVAD  LLLLATLPLM   100
           TM-1                                          TM-2

AT1   AVYTAMEYRW   PFGNHLCKIA  SASVTFNLYA  SVFLLTCLSI  DRYLAIVHPM
AT2   ATYYSYRYDW   LFGPVMCKVF  GSFLTLNMFA  SIFFITCMSV  DRYQSVIYPF   150
                               TM-3

AT1   KSRLRRTMLV   AKVTCIIIWL  MAGLASLPAV  IHRNVYFIEN  TNITVCAFHY
AT2   LSQRRNP-WQ   ASYVVPLVWC  MACLSSLPTF  YFRDVRTIEY  LGVNACIMAF   200
                     TM-4

AT1   ESRNSTLPIG   LGLT-KNILG  FLFPFLIILT  SYTLIWKALK  KAYEIQKNKP
AT2   PPEKYAQWSA   GIALMKNILG  FIIPLIFIAT  CYFGIRKHLL  KTNSYGKNRI   250
                       TM-5

AT1   RNDDIFRIIM   AIVLFFFFSW  VPHQIFTFLD  VLIQLGVIHD  CKISDIVDTA
AT2   TRDQVLKMAA   AVVLAFIICW  LPFHVLTFLD  ALTWMGIINS  CEVIAVIDLA   300
                       TM-6

AT1   MPITICIAYF   NNCLNPLFYG  FLGKKFKKYF  LQLLKYIPPK  AKSHSSLSTK
AT2   LPFAILLGFT   NSCVNPFLYC  FVGNRFQQKL  RSVFRVPITW  LQGKRETMSC   350
               TM-7

AT1   MSTLSYRPSD   NMSSSAKKPA  SCFEVE
AT2   RKSSSLREMD   TFVS                                            377
```

Fig. 2. Amino acid sequences of rat AT_{1A} and AT_2 deduced from respective cDNA. Sequences in shaded boxes are identical. Of the overall 32% sequence identity, the transmembrane regions show a somewhat higher degree of homology.

202

centrifugation we were able to isolate a cDNA insert in pCDNAI from the rat pheochromocytoma cell line PC12w [21] (Fig. 2).

The amino acid sequence of rat AT_2 deduced from the cloned cDNA was found unexpectedly to have the structural motifs of a seven membrane spanning domain receptor with 363 amino acid residues which corresponded to a theoretical molecular weight of 41,303 [21]. It also contained five potential *N*-glycosylation sites in the amino terminal extracellular domain. The difference of 30,000 from the molecular weight (71,000) of the native receptor isolated from PC12w cell plasma membrane by affinity chromatography may be accounted for by carbohydrate chains on these glycosylation sites. In AT_1 there are three potential *N*-glycosylation sites, one in the N-terminal extracellular domain and two in the second extracellular loop [12,13]. Three carbohydrate chains may account for the difference between the native molecular weight of 41,000. The comparison of the amino acid sequence of rat AT_{1A} and AT_2 revealed only 32% sequence identity, and if anything, they are only remotely related.

Functions

Although both of the isoforms turned out to be seven transmembrane domain receptors, a marked difference in their signalling and physiological functions are obvious. Their localization and regulation of expression share little commonality.

The typical and most prominent action of angiotensin II is the Gq mediated activation of PLC, which is usually followed by an opening of an L-type Ca^{2+} channel which can be blocked by its specific blockers such as dihydropyridines. It was also shown to inhibit adenylyl cyclase via a Gi mediated mechanism. Human AT_1 receptor and rat AT_{1A} receptor stably expressed in CHO cells were shown to mediate all of the three responses [22]. Since CHO cells lack endogenous angiotensin II receptor, and since only one type of angiotensin II receptor was expressed, it is clear that the AT_1 receptor can interact with three types of G proteins. Between the two AT_1 subtypes, AT_{1A} and AT_{1B} in rodents, amino acid sequences of the third cytosolic loop domain show some important diversities. This domain plays a major role in activating the G proteins. The amino acid sequences of human or bovine AT_1 more closely resembles that of rat AT_{1A} than AT_{1B}. Thus, rat AT_{1B} may have a unique feature, somewhat different from that of AT_{1A} or probably from human AT_1.

The function of AT_2 has been controversial. Using cells abundantly expressing this receptor, it has been shown not to increase $[Ca^{2+}]_i$, nor change cAMP [17-20]. However, disagreements exist as to whether angiotensin II inhibits cGMP production, and as to whether it affects phosphotyrosine phosphatase [17-20]. The cloned and stably expressed AT_2 in COS-7 cells did not show a reduction in cGMP in response to angiotensin II, nor in PC12w cells which express only AT_2 but not AT_1 [23].

Of further interest, COS-7 cells stably expressing AT_2 or PC12w cells show

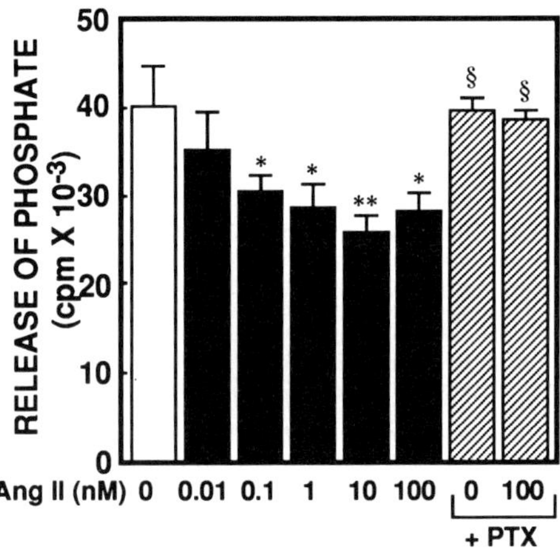

Fig. 3. Inhibition of the phosphotyrosine phosphatase activity by the AT_2 mediated action of angiotensin II in plasma membranes of COS-7 cells stably expressing AT_2. The phosphotyrosine phosphatase activity was determined with ^{32}p-Raytide™. Shaded bars indicate the reversal of angiotensin II induced inhibition of the phosphatase by pertussis toxin.

significant inhibition of phosphotyrosine phosphatase as examined by the ^{32}P-labeled peptide substrate Raytide, or p-nitrophenylphosphate [21,24] (Fig. 3). Thus, this receptor is unique in many aspects. This new group of receptors are insensitive to GTPγS with regard to ligand binding (somatostatin type 1 receptor) [21,24,25], but their functions blocked by pertussis toxin [21,24,26,27]. In contrast to other GTPγS insensitive receptors, AT_2 is unique in that it inhibits phosphotyrosine phosphatase, whereas, others activate the phosphatase. Yet, the AT_2 mediated AT_2 action seems to suppress the mitogenic action in several types of cells such as PC12w cells, COS-7 cells expressing AT_2, endothelial cells, and vascular smooth muscle cells. It is also interesting that the expression of this gene is inhibited by growth factors to varying degrees [28]. The third cytosolic loops of these PTP related receptors have some sequence homology, but the homology in this region between AT_{1A} and AT_2 are more extensive. Molecular mechanisms of AT_2 action seem unconventional and deserve further detailed analysis.

Acknowledgement

We thank Tina Stack for her editorial work. This work has been supported by research grants HL14192 and HL35323 from NIH.

References

1. Peach MJ. Biochem Pharmacol 1981;30:2745–2751.
2. Whitebread S, Mele M, Kamber B, and deGaspero M. Biochem Biophys Res Commun 1989;163:284–291.
3. Chiu AT, Herblin WF, McCall DE, Ardecky RJ, Carini DJ, Duncia JV, Pease LJ, Wong PC, Wexler RR, Johnson AL, Timmermans PBMWW. Biochem Biophys Res Commun 1989;165:196–203.
4. Chang RSL, and Lotti VJ. Mol Pharmacol 1990;29:347–351.
5. Chaki S, Inagami T. Eur J. Pharmacol 1992;225:355–356.
6. Chaki S, Inagami T. Biochem Biophys Res Commun 1992;182:388–394.
7. Chiu AT, Roscoe WA, McCall DE, Timmermans PBMWM. Receptor 1991, 1:33–40.
8. Jackson TR, Blair LAC, Marshal J, Goedert M and Hanley MR. Nature 1988, 335:437–440.
9. Rosenberg E, Kiron MAR, and Soffer R. Biochem Biophys Res Commun 1988, 151:466–472.
10. Hagiwara H, Sugiura N, Wakita K, and Hirose S. Eur J Biochem 1989, 185:405–410.
11. Seed B. Nature 1987, 329:840–842.
12. Sasaki K, Yamano Y, Bardhan S, Iwai N, Murray JJ, Hasegawa M, Matsuda Y and Inagami T. Nature 1991, 351:230–233.
13. Murphy TJ, Alexander RW, Griendling KK, Runge MS, Bernstein KE. Nature 1991, 351:233–236.
14. Iwai N, Yamano Y, Chaki S, Konishi F, Bardhan S, Tibbetts C, Sasaki K, Hasegawa M, Matsuda Y and Inagami T. Biochem Biophys Res Commun 1991;177:299–304.
15. Iwai N and Inagami T. FEBS Lett 1992;298:257–260.
16. Kakar SS, Sellers JC, Devor DC, Musgrove LC and Neill JD. Biochem Biophys Res Commun 1992;183:1090–1096.
17. Bottari SP, King IN, Reichlin S, Dahlstroem I, Lydon N and de Gasparo M. Biochem Biophys Res Comun 1992;183:206–211.
18. Purcell AG, Hodges JC, Sen I, Bumpus FM and Husain A. Endocrinology 1991;128:1947–1959.
19. Webb ML, Lieu ECK, Cohen RB, Hedberg A, Bogosian EA, Monshizadegau H, Mollogy C, Serafino R, Moreland S, Murphy TJ and Dickinson KEJ. Peptides 1992;13:499–508.
20. Leung KH, Roscoe WA, Smith RD, Timmermans PBMWM and Chiu AT. Eur J Pharmacol. 1992;227:63–70.
21. Kambayashi Y, Bardhan S, Takahashi K, Tsuzuki S, Inui H, Hamakubo T and Inagami T. J Biol Chem 1993;268:24543–24546.
22. Ohnishi J, Ishido M, Shibata T, Inagami T, Murakami K and Miyazaki H. Biochem Biophys Res. Commun 1992;186:1094–1101, 1992).
23. Speth RC, Kim KH. Biochem Biophys Res Commun 1990;169:997–1006.
24. Takahashi K, Bardhan S, Kambayashi Y, Shirai H and Inagami T. Biochem Biophys Res Commun 1994;198:60–66.
25. Rens-Domiano S, Law SF, Yamada Y, Seino S, Bell GI and Reisine T. Mol Pharmacol 1992;42:28–34.
26. Pan MG, Florio T and Stork PJS. Science 1992;256:1215–1217.
27. Florio T, Pan MMMG, Newman G, Hershberger RE, Civelli O and Stork PJS. J Biol Chem 1992;267:24169–24172.
28. Kambayashi Y, Bardhan S and Inagami T. Biochem Biophys Res Commun 1993;194:478–482.

Endothelium-Derived Factors and Vascular Functions. T. Masaki, ed.

Production of dominant negative mutations of guanylyl cyclase

Peter S.T. Yuen[2], Masahiro Takada[3], Dana K. Thompson[2,4] and David L. Garbers[1,2]

[1]*Howard Hughes Medical Institute, and* [2]*Department of Pharmacology, The University of Texas Southwestern Medical Center, Dallas, TX, U.S.A.;*
[3] *Department of Biochemistry, Tsukuba Research Laboratories, Eisai Company, Ltd., Ibaraki, Japan;*[4]*Department of Molecular Physiology and Biophysics, Vanderbilt University, Nashville, TN, U.S.A.*

Introduction

The guanylyl cyclase receptor family consists of members within both the soluble and particulate fractions of cell homogenates [1]. Four cDNA clones have been isolated for putative members of the soluble enzyme family denoted as $\alpha1$, $\alpha2$, $\beta1$, and $\beta2$ [2–7]. Only the $\alpha1\beta1$ heterodimer has been isolated from rat and bovine lung, and therefore the pairing of the remaining proteins is not known, although an enzymatically active $\alpha2\beta1$ heterodimer can be formed under conditions of coexpression [7].

The particulate forms of the family appear to contain a single transmembrane domain that separates a ligand binding domain from a protein kinase-like and a cyclase catalytic domain [1,8,9]. Currently there are four members of the family known, and they are designated as GC-A(NPR-A), GC-B(NPR-B), GC-C(STa-R) and retGC. The putative ligand for retGC has not yet been discovered, but atrial natriuretic peptide, C-type natriuretic peptide, and heat stable enterotoxins/guanylin appear to be the ligands for GC-A, GC-B and GC-C, respectively [1,9–11].

Guanylyl or adenylyl cyclases contain domains that are homologous with a region of GC-A that is known to possess catalytic activity [12]. Thus, these regions are often referred to as consensus catalytic domains, although they may in some cases possess other functions. It appears that at least two consensus catalytic domains are required for optimal expression of cyclase activity. An $\alpha1\beta1$ heterodimer possesses activity for a soluble form of guanylyl cyclase whereas expression of individual subunits results in no detectable activity [5,13,14], truncated forms of GC-A which possess enzyme activity migrate on gel filtration columns as homodimers [15], and separation of two internal consensus catalytic domains of adenylyl cyclase results in marked decreases in activity [16]. Possibly then, two consensus catalytic sites represent the minimal catalytic unit of all cyclases.

If this is the case, then it should be possible to construct mutants of the cyclases that retain their ability to dimerize or form higher-ordered structures, but lose their ability to catalyze the conversion of nucleoside triphosphates to cyclic nucleotides.

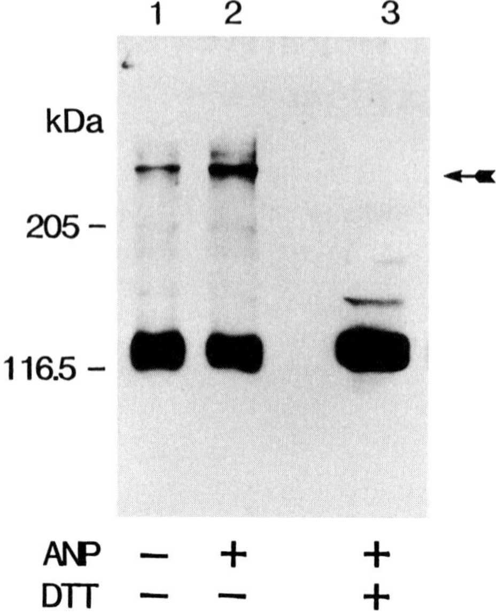

Fig. 1. The existence of a disulfide-linked GC-A dimer on SDS-PAGE. Membranes from GC-A overproducing cells were treated with or without 1 μM ANP and then electrophoresed in the presence or absence of 10 mM dithiothritol (DTT).

Such mutants could also serve as dominant negative proteins if they competed for dimerization with wild-type subunits, but in so doing blocked cyclase activity.

State of the guanylyl cyclase-A receptor

Two series of experiments were designed to determine the oligomerization state of GC-A. In the first, native GC-A was analyzed on reducing and non-reducing SDS-PAGE gels in the presence or absence of ANP/ATP, and in the second, the putative catalytic domain and a region designated as the hinge region (HCAT) was expressed and analyzed by SDS-PAGE.

In the first experiments, a small amount of immunoreactive GC-A was seen at the level of an expected homodimer in the absence of reducing agent; this material disappeared on reducing gels (Fig. 1). The relative amount of the apparent homodimer increased in the presence of ANP. Formation of the disulfide-linked, apparent homodimer of GC-A required relatively long periods of time (Fig. 2). In addition to the effects of ANP, ATP also increased formation of the apparent homodimer, and the concentrations of ATP required were approximately the same as those required for receptor activation (Fig. 3). Although there are multiple interpretations of the results, a likely possibility is that ANP/ATP induce conformational changes in GC-A that subsequently result in increased rates of disulfide exchange, since GC-A has been reported to exist as an oligomer in the absence of ANP [17,18].

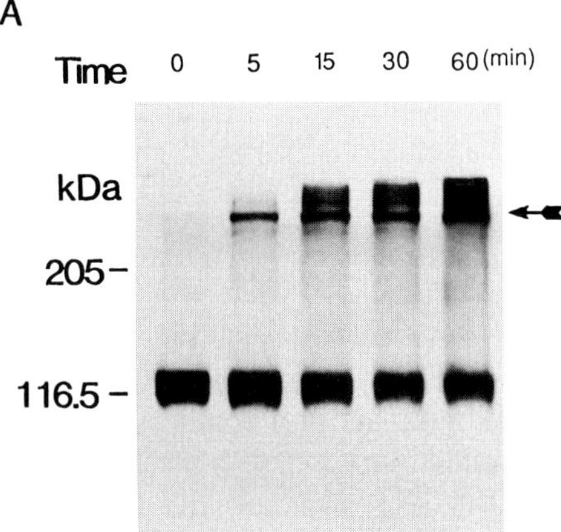

Fig. 2. Time-dependent formation of the GC-A disulfide-linked dimer after the addition of ANP. Membranes from GC-A overproducing cells were treated with 1 μM ANP for the times indicated then electrophoresed on SDS-PAGE in the absence of dithiothreitol.

In the second series of experiments, the deletion mutant HCAT was expressed in COS-7 cells and then extracts from the cells were subjected to gel permeation analysis. Fractions from such gel filtration columns were analyzed for both enzyme activity and immunoreactivity. Cyclase activity was seen at the expected position of the HCAT homodimer, and immunoreactive cyclase was observed at both the void volume and at the position of the homodimer. The addition of a chaotropic salt

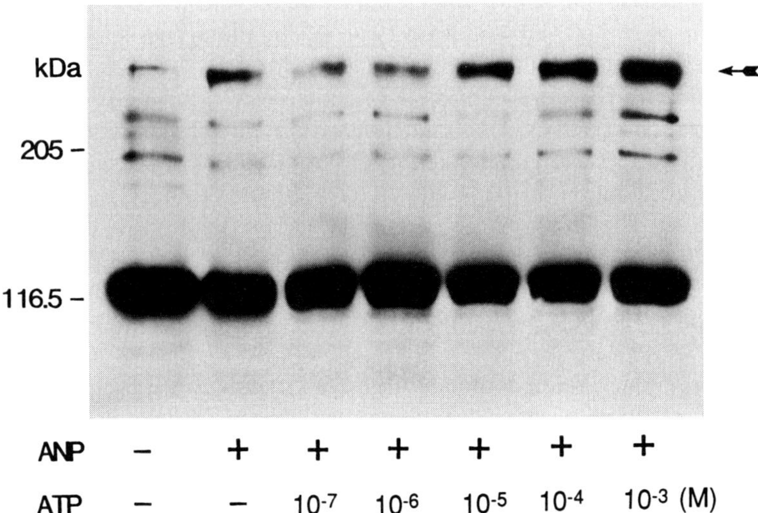

Fig. 3. The effect of ATP on formation of the disulfide-linked GC-A homodimer. Membranes of GC-A overproducing cells were treated with 1 μM ANP and the concentrations of ATP indicated on the figure. At 60 min the reactions were stopped by the addition of SDS and electrophoresed in the absence of dithiothreitol.

caused most of the immunoreactive material to now migrate as the monomer without enzyme activity.

Thus, a case can be made for an enzyme model in which dimerization is required for catalytic activity. Whether a single catalytic site is formed in response to dimerization remains unknown.

Dominant negative effect of HCAT

That HCAT formed homodimers raised the question of whether or not HCAT could form a dimer with full-length GC-A, and in so doing act in a dominant negative manner. Co-transfection of HCAT with GC-A would be expected to result in higher than normal cGMP concentrations and high guanylyl cyclase activity due to the formation of active HCAT homodimers and active HCAT/GC-A hybrid dimers. This was the case, in that basal cGMP concentrations and guanylyl cyclase activities were elevated about 45-fold and 5-fold, respectively, by cotransfection with HCAT and GC-A when compared to GC-A alone. The formation of the HCAT/GC-A hybrid, however, should result in interruption of the ANP activation signal, and this also was the case. Responses to ANP were severely blunted by the introduction of HCAT.

Alanine scanning mutagenesis of conserved residues within catalytic region

A number of highly conserved amino acid residues within the putative catalytic domain of GC-A, or of the $\alpha 1$ subunit were targeted for mutagenesis to alanine (Table 1). In COS cells that contain little if any GC-A or $\alpha 1\beta 1$, GC-A mutants or $\alpha 1$ mutants and wildtype $\beta 1$ were expressed to monitor two functions: cyclase catalytic and dimerization activities. Two mutations within the $\alpha 1$ subunit (D513A, D529A), and one within the catalytic domain of GC-A (D893A) resulted in losses of enzyme activity but retention of the ability to form dimers (Table 1).

Dominant negative effects of cyclase mutants when transfected into cells containing wild-type cyclases

The $\alpha 1$ mutations are by definition dominant negative proteins, since they resulted in undetectable cyclase activity when recombined with the wild-type $\beta 1$ subunit. However, whether the mutant α subunit would compete with wild-type α remained uncertain. For these studies, rat insulinoma cells were selected because they contained the $\alpha 1\beta 1$ form of guanylyl cyclase which responds to sodium nitroprusside. Stable cell lines were established that overproduced the various mutant forms of $\alpha 1$. The cGMP responsiveness of a rat insulinoma cell line to sodium nitroprusside, but not to ligands for the guanylyl cyclase-C receptor, was greatly diminished

Table 1. The effects of various mutations within guanylyl cyclases on enzyme activity and dimer formation. [a] Guanylyl cyclase activity was estimated in COS cells transiently transfected with the GC-A mutant alone, or in the case of α mutants, activity was estimated after co-transfection of mutant $\alpha 1$ and wild-type $\beta 1$ subunit.

Mutation	Relative cyclase activity[a]	Dimerization
$\alpha 1$ wild-type	1.0	+
$\alpha 1$ T449A	0.29	+
$\alpha 1$ D513A	N.D.	+
$\alpha 1$ K523A	0.12	+
$\alpha 1$ E525A	0.16	+
$\alpha 1$ D529A	N.D.	+
$\alpha 1$ G536A	0.34	+
$\alpha 1$ H544A	2.34	+
$\alpha 1$ E557A	0.59	+
$\alpha 1$ K614A	0.39	+
$\alpha 1$ R623A	2.20	+
GC-A D877A	N.D.	−[*]
GC-A K887A	N.D.	−[*]
GC-A D893A	N.D.	+
GC-A G900A	N.D.	−[*]
GC-A H909A	N.D.	−[*]
GC-A R940A	N.D.	−[*]
GC-A H944A	N.D.	−[*]
GC-A E974A	30	+

[*]These mutant proteins formed high molecular weight aggregates that may represent denatured protein.

by the presence of either dominant negative $\alpha 1$ subunit, suggesting that overexpression specifically inhibits the nitric oxide/cGMP signaling pathway.

The GC-A mutants, although lacking cyclase activity and retaining the ability to dimerize, would not necessarily act as dominant negative proteins. To assess whether dominant negative effects would be seen, the mutant protein and wild-type GC-A were co-transfected into COS cells. When cells were transfected with GC-A alone, typical ANP response curves were obtained with one-half maximal stimulation of cyclic GMP elevations at 1 nM and maximum stimulation at about 100 nM. Upon co-transfection with the point mutant, however, cyclic GMP responses were greatly reduced despite the expression of equivalent amounts of wild-type GC-A.

Conclusions

Specific inhibitors of the various forms of guanylyl cyclase have not been available but would facilitate studies on the functions of cyclic GMP in cells. Here, we describe various mutations of either a particulate cyclase, GC-A, or of a soluble cyclase ($\alpha 1 \beta 1$) that result in the formation of dominant negative proteins. Introduction of the proteins into cells can result in a marked and specific inhibition of ligand-dependent cyclic GMP responses.

Acknowledgements

This work was supported in part by grant I1233 from the Welch Foundation.

References

1. Garbers DL. Cell 1992;71:1–4.
2. Koesling D, Herz J, Gausepohl H, Niroomand F, Hinsch KD, Mulsch A, Bohme E, Schultz G, Frank R. FEBS Lett 1988;239:29–34.
3. Nakane M, Saheki S, Kuno T, Ishii K, Murad F. Biochem Biophys Res Comm 1988;157:1139–1147.
4. Koesling D, Harteneck C, Humbert P, Bosserhoff A, Frank R, Schultz G, Bohme E. FEBS Lett 1990;266:128–132.
5. Nakane M, Arai K, Saheki S, Kuno T, Buechler W, Murad F. J Biol Chem 1990;265:16841–16845.
6. Yuen PST, Potter LR, Garbers DL. Biochemistry 1990;29:10872–10878.
7. Harteneck C, Wedel B, Koesling D, Malkewitz J, Bohme E, Schultz G. FEBS Lett 1991;292:217–222.
8. Chinkers M, Garbers DL, Chang MS, Lowe DG, Chin H, Goeddel DV, Schulz S. Nature 1989;338:78–83.
9. Schulz S, Singh S, Bellet RA, Singh G, Tubb DJ, Chin H, Garbers DL. Cell 1989;58:1155–1162.
10. Schulz S, Green CK, Yuen PST, Garbers DL. Cell 1990;63:941–948.
11. Currie MG, Fok KF, Kato J, Moore RJ, Hamra FK, Duffin KL, Smith CE. Proc Natl Acad Sci USA 1992;89:947–951.
12. Thorpe DS, Morkin E. J Biol Chem 1990;265:14717–14720.
13. Harteneck C, Koesling D, Soling A, Schultz G, Bohme E. FEBS Lett 1990;272:221–223.
14. Buechler W, Nakane M, Murad F. Biochem Biophys Res Comm 1991;292:351–357.
15. Thorpe DS, Niu S, Morkin E. Biochem Biophys Res Comm 1991;180:538–544.
16. Tang WJ, Krupinsky J, Gilman AG. J Biol Chem 1991;266:8595–8603.
17. Lowe DG. Biochemistry 1992;31:10421–10425.
18. Chinkers M, Wilson EM. J Biol Chem 1992;267:18589–18597.

Endothelium-Derived Factors and Vascular Functions. T. Masaki, ed.

Regulation of receptor-G protein signaling by the effector enzyme phospholipase C-β

Elliott M. Ross, Gabriel Berstein and Mahesh Karandikar
Department of Pharmacology, University of Texas, Southwestern Graduate School of Biomedical Sciences, 5323 Harry Hines Blvd., Dallas, TX 75235-9041, U.S.A.

Introduction

Numerous endocrine and paracrine agents that regulate the vascular endothelium and the circulating cells therein act through G protein-coupled receptors. This group includes peptides such as endothelin and angiotensin, eicosanoids and biogenic amines (adrenergics, serotonin, and the muscarinic effects of acetylcholine). Many have been discussed at this symposium. The G proteins organize information from the receptors and in turn regulate cellular function through a multitude of second messenger pathways: release of IP_3 to elevate Ca^{2+} levels, activation of K^+ channels, inhibition and activation of adenylyl cyclase, etc. The diversity of these signals derives from the receptors' selective use of different G proteins to regulate distinct cytoplasmic effector proteins, which are enzymes or channels that synthesize, degrade or release second messengers.

This pattern of signal sorting is perhaps best exemplified by the five muscarinic cholinergic receptors. Ashkenazi and Peralta [1,2] demonstrated that M1 (and M3 and M5) receptors principally utilize a pertussis toxin-insensitive G protein to activate a phosphatidylinositol-diphosphate-specific phospholipase C (a PLC-β [3]). They also act with much lower efficiency through a toxin-sensitive G protein in some cells. In contrast, they showed that M2 (and M4) receptors act through toxin-sensitive G proteins, probably G_i's, which characteristically mediate opening of the M-type K^+ channel, inhibition of adenylyl cyclase and activation of PLC-β (more weakly, probably via the βγ subunits) [4].

When we initially expressed and purified recombinant M1 and M2 muscarinic receptors using baculovirus vectors, we found the M2 receptor efficiently activates the G_i's, G_o and G_z [5]. Activation of G_z was unexpected, but may explain certain toxin-insensitive effects of the M2 or M4 receptor in the nervous system, where G_z is primarily expressed. The M1 receptor regulated none of these G proteins, nor did it regulate G_s. It should be remembered that coupling of a receptor to a G protein that it would not normally regulate in vivo can be driven either by reconstituting the G protein with high concentrations of the receptor or by vast overexpression of the receptor in transfected cells. This potential derives from the homology shared within

212

the receptor and G protein families. However, the results referred to here clearly showed that the M2 receptors display a marked selectivity for G_i's, G_o, and G_z and that the M1 receptors did not regulate these G proteins even at high concentrations.

The G_q family of G proteins, initially identified by cDNA cloning in Simon's laboratory [6,7], were likely candidates for mediators of the M1 signaling pathway. They are pertussis toxin-insensitive, and both G_q and G_{11} (a member of the G_q family) were shown by the groups of Sternweis and Exton to stimulate PLC-β1 [8,9]. We used in vitro biochemical approaches to show the M1 muscarinic regulates PLC-β1 through the G_q family of G proteins, and these studies led to the unexpected finding that PLC-β substantially increases the rate of GTP hydrolysis and consequent deactivation of G_q. Details of the initial studies are provided elsewhere [10,11]; some of their implications and more recent corollaries are discussed below.

Methods

The basic reconstituted system used in the experiments described here consists of phospholipid vesicles that contain human M1 muscarinic receptor purified from baculovirus-infected Sf9 cells [5] and $G_{q/11}$, an approximately equimolar mixture of G_q and G_{11} purified from bovine liver [12]. Key results have now been confirmed with recombinant G_q and $G\beta\gamma$ purified from Sf9 cells infected with the appropriate baculoviruses, generously provided by John Hepler [G_q) or Jorge Iñiguez-Lluhi ($G\beta\gamma$) (both of this department). PLC-β1, purified from either bovine brain [13,14], HeLa cells infected with recombinant vaccinia virus [15] or Sf9 cells infected with recombinant baculovirus, was added to the vesicles after reconstitution. All relevant procedures have been described in detail elsewhere [10,11].

Drosophila melanogaster were grown on commercial medium. Visually wild-type white eye (w[1118]) flies were of the Oregon Red strain. NorpA[EE5] flies also carried the white eye marker. Crude membrane fractions were prepared from whole heads. Tissue was removed with a razor blade and allowed to dark adapt for 60 min at 0°C in homogenization buffer (10 mM Mops (pH 7.0), 250 mM sucrose, 120 mM KCl, 5 mM $MgCl_2$, 1 mM dithiothreitol, 0.5 mM phenylmethylsulfonylfluoride, 10 µg/ml leupeptin and 1 µg/ml pepstatin A [16,17]). Further procedures were performed under dim red light. Tissue was homogenized in a tight-fitting ground glass homogenizer. In a few cases, membranes were collected by centrifugation before assay, but results with membranes and whole homogenates did not differ noticeably. GTPγS binding was measured after 5 min at 15°C in buffer that contained 50 mM Mops (pH 6.7), 5 mM $MgCl_2$, 0.5 mM App(NH)p, 20 nM [^{35}S]GTPγS, 1 mM dithiothreitol and 0.2 mg/ml bovine serum albumin using the general assay procedures described previously [18]. While not at equilibrium, this combination of time and conditions gives an optimal and reproducible stimulation by light. GTPase activity was measured at 15°C in buffer that contained 40 mM Mops (pH 6.7), 0.1 µM [γ-^{32}P]GTP, 6 mM $MgCl_2$, 1 mM EDTA, 1 mM App(NH)p, 0.5 mM ATP, 3 mM K_2-

phospho-enol-pyruvate, 10 µg/ml pyruvate kinase, 1 mM dithiothreitol and 0.2 mg/ml bovine serum albumin [19], also as described previously [18]. Both assays were conducted either in the dark or under continuous blue light and following 1 min exposure to blue light. The light source was either a photographic strobe lamp or a 100 W tungsten lamp with a fiber optic conductor. Bleaching light was filtered through a Schott BG28 blue filter.

Results and Discussion

G proteins mediate signaling by traversing a controlled cycle of activation and deactivation. G proteins are activated when they bind GTP, such that they can in turn activate cellular effector proteins such as PLC-β. Deactivation and consequent termination of signaling occur when bound GTP is hydrolyzed to GDP. Receptors act by catalyzing the exchange of bound GDP for GTP (Fig. 1, see ref. [20]). G_q is somewhat unusual among G proteins because its functional activation by GTP or GTPγS is difficult to demonstrate after purification, as is the binding of radio-labeled GTPγS. For unknown reasons, activation by GTPγS is significant only after prolonged incubation with unusually high GTPγS concentrations, and measurement of the binding of radiolabeled GTPγS to isolated G_q has only been estimated recently [21]. In fact, initial identification of G_q as a PLC activator depended on its activation by F^- plus Al^{+3} [8,9].

In initial experiments using phospholipid vesicles reconstituted to contain receptors and G_q, we found that agonist-liganded M1 muscarinic receptor efficiently catalyzed the binding of GTPγS to G_q and its consequent activation. When we compared activation rates in vesicles that contained no receptor, M1 receptor, or M2 receptor, we found that even the unliganded M1 receptor significantly accelerated G_q activation. The agonist-liganded M2 receptor stimulated GTPγS binding to G_q only slightly, at about the same rate as did the unliganded or antagonist-liganded

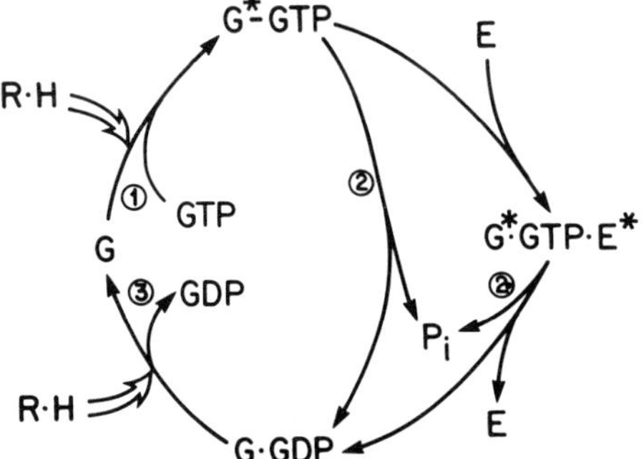

Fig. 1. A G protein (G) is activated by GTP binding (reaction 1), and activated G protein can bind and activate its effector (E). GTP hydrolysis terminates activation (reaction 2). PLC-β accelerates the hydrolysis of GTP bound to G_q, at least 50-fold (reaction 2·). Reactivation requires the release of bound GDP and the binding of GTP (reactions 3 and 1), and is catalyzed by receptors (from ref. [20]).

214

M1 receptor. This finding is consistent with the predominantly pertussis toxin-sensitive action of M2 receptors in cells [4]. As expected from previous studies [8,9], GTPγS-liganded G_q stimulated PLC-β1 activity at least 40-fold.

G protein-mediated signaling in vivo is determined both by the rate of receptor-catalyzed GDP/GTP exchange and by the rate of hydrolysis of bound GTP. The M1 receptor efficiently catalyzed both the binding of GTP and the dissociation of GDP by G_q, which resulted in substantial (greater than 10-fold) stimulation of the steady-state GTPase activity [10]. Because G_q hydrolyzes bound GTP slowly, with a $t_{1/2}$ of about 1 min, receptor-catalyzed GTP/GDP exchange results in the steady-state accumulation of a relatively large amount of the activated species, G_q-GTP (Fig. 2). The steady-state level of activated G_q was consistent with the level predicted by calculations based on the individually determined rate constants for the partial reactions shown in Fig. 1 [10].

We then used GTPγS-activated $G_{q,11}$ to estimate how much GTP-liganded G_q should be necessary to observe a given level of PLC-β1 activation in vesicles that contain receptor, G_q and PLC. Calculations based on these data and data of the sort shown in Fig. 2 suggested that carbachol added to the vesicles should substantially activate PLC-β1 in the presence of GTP. However, we were unable to demonstrate GTP-supported PLC stimulation by carbachol in the receptor-G_q-PLC system. Failure to obtain an expected result in a complex reconstituted system is always treacherous to interpret, but a possible explanation was that PLC-β1 was stimulating the hydrolysis of G_q-bound GTP, such that our initial estimates of its steady-state accumulation would be in error. This provocative explanation turned out to be correct.

As shown in Fig. 3, PLC-β1 stimulates the steady-state GTPase activity of M1

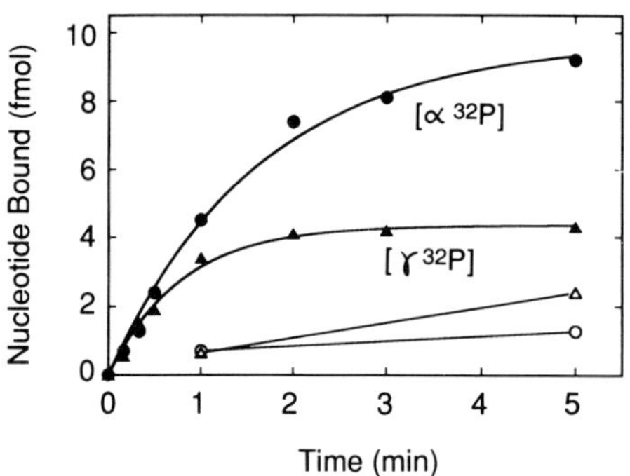

Fig. 2. The binding of GTP to $G_{q/11}$ is stimulated by the M1 muscarinic receptor. Binding and hydrolysis of GTP can be assayed using GTP labeled at either the α or γ positions with ^{32}P. Bound [γ-^{32}P]GTP (triangles) indicates the concentration of G_q bound to GTP only. Bound [γ-^{32}P]GTP (circles) reflects both GTP-bound G_q and G_q that has already hydrolyzed its bound GTP to GDP. The steady-state level of bound [γ-^{32}P]GTP indicates the amount of activated G (from ref. [10]).

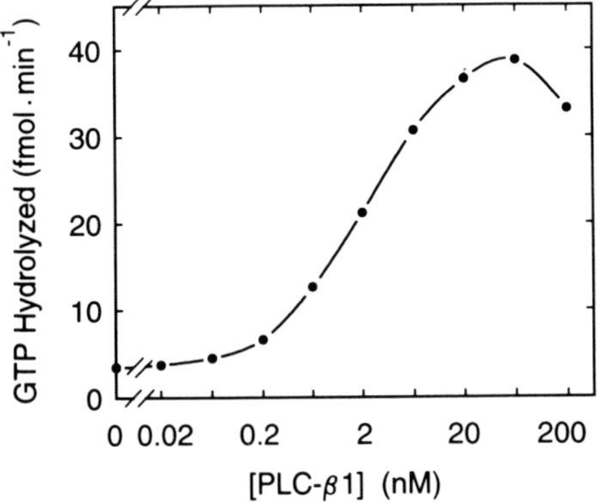

Fig. 3. PLC-β 1 stimulates the GTPase activity of $G_{q/11}$. The steady-state GTPase activity of M1 receptor-$G_{q/11}$ vesicles was measured in the presence of 1 mM carbachol and increasing concentrations of purified, recombinant PLC-β 1 (from ref. [11]).

receptor-G_q vesicles. About 20-fold stimulation is shown here, and 40-fold stimulation can be observed. Numerous control experiments indicated that PLC-β1 regulates G_q directly. For example, several different monoclonal antibodies specific for different epitopes on PLC-β1 blocked both its stimulation of the G_q-catalyzed GTPase reaction and its own stimulation by activated G_q with the same rank order of potency. Thus, as documented more fully elsewhere [11], PLC-β1 acts as a GTPase activating protein, or GAP, for its physiologic regulator G_q. The EC_{50} for PLC-β1 as a G_q GAP is 1-2 nM (Fig. 3), about equal to the EC_{50} with which GTPγS-activated G_q stimulates PLC-β1 [10], reflecting the K_d for the binding of the two proteins to each other.

The GAP activity of PLC-β1 is specific for G_q, its physiological regulator. It had no effect on the GTPase activity of G_s, G_i, or G_o. Conversely, PLC-c (which is not G protein-regulated) displayed no G_q-GAP activity. A sample of purified recombinant ras-GAP (a gift from Dr. Frank McCormick, Onyx) also did not display G_q-GAP activity, although its regulatory activity with purified p21ras was not assayed in parallel at the time of this experiment.

To test whether PLC-β1 displays significant G_q-GAP activity in intact biological membranes as well as in reconstituted systems, we took advantage of the *Drosophila* photoreceptor, in which rhodopsin stimulates GTP binding to a G_q. This G_q in turn activates a PLC-β1 to generate a cytoplasmic Ca^{2+} signal [22]. *Drosophila* is particularly attractive as an experimental system because there exist mutants in many of the key protein components of the signaling pathway. We asked whether the loss of NorpA, the photoreceptor-specific PLC-β, would influence the GTPase activity of the visual G_q when it is stimulated to bind GTP by rhodopsin. We used crude membrane and homogenate fractions prepared from either eyes or heads and measured light-stimulated GTPγS binding and GTP hydrolysis. Because there is considerable background nucleotide binding and nucleoside triphosphatase activity in these membranes that is unrelated to the phototransduction cascade, experiments

216

were performed in the presence of both ATP and App(NH)p, and results are expressed as the relative incremental activities caused by exposure of the membranes to light. Illumination of wild-type *Drosophila* photoreceptor membranes caused an increase both in the rate of binding of GTPγS and of the steady-state GTPase activity. These activities are comparable to those obtained in the more detailed studies of *Drosophila* and *Musca* by Selinger, Minke and co-workers, although activities in *Drosophila* membranes are smaller than those observed in *Musca* membranes [17]. GTPase and GTPγS binding activities should not be taken as absolute measures of the capacity of the system because the assay conditions are adjusted to suppress high levels of non-specific nucleotide triphosphatases and nucleotide-binding proteins.

To determine the contribution of a GAP activity of the NorpA PLC-β to the steady state GTPase, we compared the activities of head homogenates from wild-type and norpA mutants. As shown in Table 1, norpA membranes displayed about half the GTPγS binding activity of wild-type membranes, which indicates that rhodopsin and G_q are both present at near normal levels and can interact productively in norpA eyes. In contrast, norpA membranes did not display any steady-state GTPase activity. Activity was not significantly above the dark control in any of seven experiments. Thus, the NorpA PLC-β1 appears to act as a GAP for G_q in the *Drosophila* visual pathway.

The finding that PLC is a GAP for G_q [11] and the simultaneous finding that the retinal cyclic GMP phosphodiesterase may act as a GAP for transducin [23] are the first demonstrations of the existence of GAP's for trimeric G proteins. At least in the case of the G_q-GAP activity of PLC-β1, the effect appears to be primarily a stimulation of the rate of hydrolysis of bound GTP. Thus, stimulation of steady-state GTPase activity of G_q is dependent on receptor-catalyzed GTP/GDP exchange because basal exchange rates are quite low. In addition, when we attempted to measure the rate of hydrolysis of bound GTP directly, we found that a substantial fraction (20-40%) of bound GTP was hydrolyzed extremely rapidly, at least 50-fold faster than the rate observed in the absence of PLC. The G_q GAP activity thus appears similar to the activity expressed by GAP's for the small, monomeric GTP-binding proteins such as p21ras.

Using G protein-regulated effectors as GAP's for their specific G protein regulators is an attractive regulatory mechanism. It allows different effectors to display distinctive temporal patterns of signaling. Because GTP hydrolysis is a major determinant of a signal's amplitude, it also allows effectors to display

Table 1.

	GTPγS Binding	GTPase
WT	91 ± 14 (*n* = 4)	18.6 ± 7.2 (*n* = 8)
norpA	44 ± 4 (*n* = 4)	1.4 ± 2.9 (*n* = 7)

All data are light-stimulated activities, expressed as per cent increases over parallel dark controls, ± standard deviation.

diversity in the intensity of their responses to stimulation by a single class of G protein. It assigns a unique quantal output to each effector. Thus, a single G protein could activate multiple effectors with uniquely appropriate time constants and/or efficiencies. Such individuality of signaling can be enhanced further if allosteric or covalent modification of the effector alters its GAP activity.

It is likely that other G protein-regulated effectors are also GAP's for their respective G protein activators. Comparison of the rapid in vivo deactivation rates of G protein-gated ion channels with the much slower rates of GTP hydrolysis by pure G proteins in vitro suggests that the channels will also be GAP's [24]. In contrast, no G_s GAP activity has been observed for adenylyl cyclase, either in early reconstitution experiments [25] or in more recent attempts. Adenylyl cyclase may lack G_s GAP activity, and the rate of hydrolysis of G_s-bound GTP is roughly equivalent to the rate of deactivation of adenylyl cyclase upon removal of agonist (see [26] and references therein). In contrast, we predict that the G_s-gated Ca^{2+} channel will be a G_s GAP because it is very rapidly deactivated.

The GAP activity of effectors may also enhance the specificity with which a single receptor directs its signals to different G protein-effector pathways. Preliminary kinetic studies of the M1 muscarinic receptor-G_q-PLC-$\beta1$ system suggest that the relevant signal processor may be a receptor-G_q-effector complex [11,27]. This complex should be stable only in the presence of agonist and GTP, and it probably has a lifetime of perhaps 10-50 catalytic cycles. Its formation has the capacity to fine-tune the ability of a G protein to select among homologous receptors according to their abilities to form the complex. This idea explains why receptors that are quite selective for a specific signaling pathway in cells are far less specific for the relevant G proteins in vitro. Preliminary observations indicate that different receptors that seem to stimulate the activation of G_q with nearly identical efficiencies can differ dramatically when PLC-$\beta1$ is present.

Acknowledgements

The initial work on the G_q-GAP activity of PLC-$\beta1$ was a collaboration with Jonathan Blank and John Exton (Vanderbilt University) and with Dock-Young Jhon, Dongeun Park and Sue Goo Rhee (NHLBI). We are indebted to colleagues who gave us absolutely essential help, advice and flies for the experiments on the dG_q/NorpA system: Flora Katz, Dennis McKearin, Baruch Minke, William Pak, Zvi Selinger, Dean Smith, Steve Wassermann and Charles Zucker. Experiments performed in our laboratory were supported by NIH grant R37GM30355 and fellowship S32GM14489 and R.A. Welch Foundation grant I-0982.

218

References

1. Ashkenazi A, Peralta EG, Winslow JW, Ramachandran J, Capon DJ. Cell 1989;56:487–493.
2. Ashkenazi A, Winslow JW, Peralta EG, Peterson GL, Schimerlik MI, Capon DJ, Ramachandran J. Science 1987;238:672–675.
3. Rhee SG, Choi KD. J Biol Chem 1992;267:12393–12396.
4. Hulme EC, Birdsall NJM, Buckley NJ. Annu Rev Pharmacol Toxicol 1991;30:633–673.
5. Parker EM, Kameyama K, Higashijima T, Ross EM. J Biol Chem 1991;266:519–527.
6. Strathmann M, Simon MI. Proc Natl Acad Sci USA 1990;87:9113–9117.
7. Strathmann M, Wilkie TM, Simon MI. Proc Natl Acad Sci USA 1989;86:7407–7409.
8. Smrcka AV, Hepler JR, Brown KO, Sternweis PC. Science 1991;251:804–807.
9. Taylor SJ, Chae HZ, Rhee SG, Exton JH. Nature 1991;350:516–518.
10. Berstein G, Blank JL, Smrcka AV, Higashijima T, Sternweis PC, Exton JH, Ross EM. J Biol Chem 1992;267:8081–8088.
11. Berstein G, Blank JL, Jhon D-Y, Exton JH, Rhee SG, Ross EM. Cell 1992;70:411–418.
12. Blank JL, Ross AH, Exton JH. J Biol Chem 1991;266:18206–18216.
13. Ryu SH, Cho KS, Lee K-Y, Suh P-G, Rhee SG. J Biol Chem 262:12511–12518.
14. Ryu SH, Suh P-G, Cho KS, Lee K-Y, Rhee SG. Proc Natl Acad Sci USA 1987;1987;84:6649–6653.
15. Park D, Jhon D-Y, Kriz R, Knopf J, Rhee SG. J Biol Chem 1992;267:16048–16055.
16. Byk T, Bar-Yaacov M, Doza YN, Minke B, Zelinger Z. Proc Natl Acad Sci USA 1993;90:1907–1911.
17. Devary O, Heichl O, Blumenfeld A, Cassel D, Suss E, Barash S, Rubinstein CT, Minke B, Selinger Z. Proc Natl Acad Sci USA 1987;84:6939–6943.
18. Brandt DR, Ross EM. J Biol Chem 1986;261:1656–1664.
19. Blumenfeld A, Erusalimsky J, Heichal O, Selinger Z, Minke B. Proc Natl Acad Sci USA 1985;82:7116–7120.
20. Ross EM. Neuron 1989;3:141–152.
21. Hepler JR, Kozasa T, Smrcka AV, Simon MI, Rhee SG, Sternweis PC, Gilman AG. J Biol Chem 1993;268:14367–14375.
22. Minke B, Selinger Z. In: Osborne NN, Chader GJ (eds.) Progress in Retinal Research 11. Oxford: Pergamon, 1992;100–124.
23. Arshavsky VY, Bownds MD. Nature 1992;357:416–417.
24. Breitwieser GE, Szabo G. Nature 1985;317:538–540.
25. May DC, Ross EM, Gilman AG, Smigel MD. J Biol Chem 1985;260:15829–15833.
26. Brandt DR, Asano T, Pedersen SE, Ross EM. Biochemistry 1983;22:4357–4362.
27. Ross EM. Rec Prog Hormone Res 1994(50) in press.

Endothelium-Derived Factors and Vascular Functions. T. Masaki, ed.

Phenotypic change of endothelin receptor subtype in vascular smooth muscle cells: its implication in atherosclerosis

Yukio Hirata and Satoru Eguchi

Endocrine-Hypertension Division, Second Department of Internal Medicine, Tokyo Medical and Dental University, Yushima 1-5-45, Bunkyo-ku, Tokyo 113, Japan

Summary

Vascular ET_A receptors are functionally coupled to different G-proteins, such as Gq and Gs, thereby activating phospholipase C and adenylate cyclase, respectively. Phenotypic change of vascular smooth muscle cell is concomitantly associated with receptor subtype change from ET_A to ET_B as well as overexpression of ET-1 gene. These data suggest that phenotypic change of ET receptor subtype change associated with overexpression of ET-1 gene by vascular smooth muscle cells may be involved in vascular remodeling.

Introduction

Endothelin (ET)-1 derived from vascular endothelial cells (EC) is a potent constrictor [1] and mitogen for vascular smooth muscle cells (VSMC) [2]. Two distinct ET receptor subtypes, designated ET_A and ET_B, have been cloned and sequenced [3,4]; ET_A receptors show selective affinity for ET-1, whereas ET_B receptors show nonselective affinity for three isopeptides (ET-1, ET-2, ET-3). Both ET_A and ET_B are G-protein-coupled receptors and functionally linked to phospholipase (PL) C to induce phosphoinositide breakdown [3–6]. It has been shown that VSMCs predominantly possess ET_A receptors that mediate vasoconstriction, while ECs predominantly possess ET_B receptors that mediate vasodilation via generation of nitric oxide and prostacyclin [5,6].

VSMCs undergoes phenotypic modulation from contractile to synthetic phenotype during in vitro culture conditions. Cultured VSMCs with synthetic phenotype resemble those in the neointima of atherosclerosis. However, the questions remain unanswered as to whether ET_A receptors in VSMC are functionally coupled to multiple G-proteins and signal transduction systems, and whether ET receptor subtype phenotypically changes during in vitro culture conditions, and if so, whether phenotypic change is associated with enhanced mitogenic effect of ET-1 as an autocrine/paracrine factor.

In the present communication, to address these questions we will discuss ET and ET receptor subtype in rat aortic VSMC mainly obtained from our laboratory.

ET$_A$ receptor subtype and different G-proteins

Northern blot analysis of RNA from rat aortic media and cultured VSMC from the early passage with cDNA for rat ET$_A$ receptor as a probe revealed two hybridization bands (5.2 and 4.2 kb) corresponding to the sizes of rat ET$_A$ mRNAs (Fig. 1). Binding studies of rat VSMCs also showed that ET$_A$ receptor antagonist (BQ-123) has higher binding affinity than ET$_B$ receptor agonist (BQ-3020), indicating the predominant expression of ET$_A$ receptor in rat aortic VSMC [7]. ET-1 dose-dependently (10^{-9}–10^{-7} M) stimulated inositol 1,4,5-trisphosphate (IP$_3$) formation in cultured rat VSMC, whose effect was blocked by BQ-123 with approximate IC$_{50}$ of 10^{-8} M [5]. These data are consistent with the notion that ET$_A$ receptor subtype in rat VSMC is functionally linked to PLC via Gq protein to induce phosphoinositide breakdown.

ET-1 also dose-dependently (10^{-9}–10^{-6} M) stimulated cyclic AMP (cAMP) formation in cultured rat VSMC, whose effect was similarly blocked by BQ-123 with the approximate IC$_{50}$ of 10^{-8} M. The ET-1-stimulated cAMP formation was unaffected by cyclo-oxygenase inhibitor (indomethacin) or PLA$_2$ inhibitor (quinacrine), but additive with isoproterenol [8]. These data suggest that ET$_A$ receptor subtype in rat VSMC is also coupled to adenylate cyclase (AC) via Gs protein to generate cAMP, which is distinct from β-adrenergic receptor or arachidonate cascade. To ascertain that rat VSMCs have Gs proteins, ADP-ribosylation of VSMC membrane by cholera toxin was performed. Cholera toxin specifically ADP-ribosylated 45- and 52-kDa proteins, whose molecular sizes are similar to those of α-subunits of Gs proteins of vascular smooth muscle [8]. Taken together, our data suggest that, in addition to PLC via Gq, ET$_A$ receptors are functionally coupled to AC via Gs in rat VSMC. The physiological role of ET$_A$ receptor-mediated cAMP formation in VSMC remains open to questions.

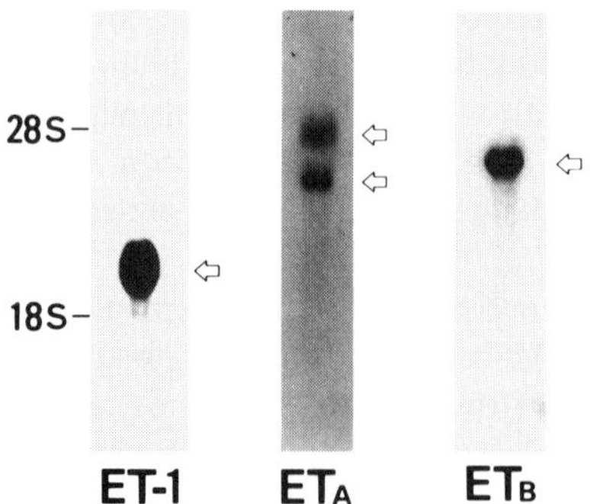

Fig. 1. Expression of ET-1 (left), ET$_A$ receptors (middle) and ET$_B$ receptor (right) mRNAs by cultured rat vascular smooth muscle cells. Total RNAs from the early passage VSMC (ET$_A$) and the late passage VSMC (ET$_B$) were hybridized with cDNAs for rat ET$_A$ and ET$_B$ receptor cDNAs as probes, respectively. Poly(A+) selected RNA from the late passage VSMC was hybridized with rat ET-1 cDNA as a probe. Arrows indicate mRNA sizes of ET-1 (2.3 kb), ET$_A$ (5.2 and 4.2 kb) and ET$_B$ (4.8 kb), respectively.

Phenotypic change of ET receptor subtype

To determine whether vascular ET receptor subtypes phenotypically changes during in vitro culture conditions, we studied the expression of ET_A and ET_B receptor subtypes as evaluated by binding studies and Northern blot analysis, and DNA synthesis as measured by [³H]thymidine uptake in cultured rat VSMCs during serial passages [7]. Binding studies using [¹²⁵I]ET-1 as a radioligand revealed that the early passage (5th-10th) VSMCs had binding affinities for ET isopeptides characteristic for ET_A receptors (ET-1>BQ-123>>BQ-3020), whereas the late passage (30th-35th) VSMCs had binding affinities in the rank order of ET-1>>BQ-123=BQ-3020=ET-3. However, Scatchard analysis of binding data from the early and the late passage VSMC showed the presence of binding sites with the comparable K_d (~10^{-10} M), but the maximal binding capacity in the late passage (76,000 sites/cell) was about twice as that of the early passage (34,000 site/cell). Binding studies of the late passage VSMC in the presence of excess of BQ-123 to fully block ET_A receptors revealed that ET-1, ET-3 and ET_B receptor agonist (BQ-3020) had almost the same binding affinity. In fact, nonselective ET_A/ET_B receptor antagonist ([Thr¹⁸, γ-MeLeu¹⁹]ET-1) had the same binding affinity to that of ET-1, while ET_A receptor antagonist (BQ-123) had far less affinity in the late passage VSMC. These data suggest that ET_A receptors predominate in the early passage VSMC, while ET_B receptors, in addition to ET_A receptors, are predominant in the late passage VSMC.

Northern blot analysis of RNA from rat VSMC from the early and the late passage with cDNAs for ET_A and ET_B receptors as probes also revealed that ET_A receptor mRNA was more predominant in the early passage, whereas ET_B receptor mRNA was more abundant in the late passage (Fig. 2), thus verifying the results obtained from binding studies.

Potentiation of mitogenic effect by ET-1

ET-1 stimulated DNA synthesis (about 2.5-fold) in the early passage VSMC, but remarkably (about 10-fold) stimulated DNA synthesis in the late passage VSMC [7].

Fig. 2. Immunohistochemical localization of ET-1 in human coronary artery. Abitin-biotin peroxidase methods using rabbit polyclonal ET-1 antibody was performed. Note ET-1-like immunoreactivity in vascular endothelial cells and modulated vascular smooth muscle cells in the neointima.

4. Sakurai T, Yanagisawa M, Takuwa Y, Miyazaki H, Kimura S, Goto K, Masaki T. Nature 1990; 348:732–735.
5. Eguchi S, Hirata Y, Ihara M, Yano M, Marumo F. FEBS Lett 1992;302:243–246.
6. Hirata Y, Emori T, Eguchi S, Kanno K, Imai T, Ohta K, Marumo F. J Clin Invest 1993;91: 1367–1373.
7. Eguchi S, Hirata Y, Imai T, Kanno K, Marumo F. Endocrinology 1994;134:222–228.
8. Eguchi S, Hirata Y, Imai T, Marumo F. Endocrinology 1993;132:524–529.
9. Shichiri M, Hirata Y, Ando K, Emori T, Ohta K, Kimoto S, Ogura M, Inoue A, Marumo F. Hypertension 1990;15:493–496.
10. Kanno K, Hirata Y, Numano F, Emori T, Ohta K, Shichiri M, Marumo F. JAMA 1990;264:28–68.
11. Lerman A, Edwards BS, Hallett JW, Heublein DM, Sandberg SM, Burnett JC. N Engl J Med 1991;325:997–1001.

Adhesion molecules

Endothelium-Derived Factors and Vascular Functions. T. Masaki, ed.

Regulation of P-selectin function and expression

Rodger P. McEver

W.K. Warren Medical Research Institute, Departments of Medicine and Biochemistry, University of Oklahoma Health Sciences Center, and, Cardiovascular Biology Research Program, Oklahoma Medical Research Foundation, Oklahoma City, OK 73104 U.S.A.

Summary

The selectins are three related Ca^{2+}-dependent lectins that initiate leukocyte adhesion to the blood vessel wall in response to inflammatory stimuli. P-selectin, expressed by activated platelets and endothelium, binds with high affinity to a limited number of protease-sensitive sites on myeloid cells. These sites may correspond to a specific sialoglycoprotein ligand for P-selectin in myeloid cell extracts that may play an important role in cell adhesion. The 5′ flanking region of the P-selectin gene contains elements that direct its cell type-specific expression and may increase transcription in response to inflammatory cytokines. The surface expression of P-selectin is further controlled by signals in the cytoplasmic domain that direct, respectively, sorting into secretory granules, endocytosis, and movement from endosomes into lysosomes. Dysregulated surface expression of P-selectin may contribute to the pathogenesis of inflammatory and thrombotic disorders.

Introduction

The selectins are a group of three related membrane proteins that mediate the initial rolling of leukocytes on the blood vessel wall in response to tissue injury or infection. L-selectin is constitutively expressed on leukocytes and binds to ligands on high endothelial venules of lymph nodes and to inducible ligands on activated endothelium at sites of inflammation. E-selectin is expressed by cytokine-stimulated endothelium, where it mediates adhesion of leukocytes. P-selectin is expressed by activated platelets and endothelial cells, where it also mediates leukocyte adhesion. The transient adhesion mediated by the selectins allows rolling leukocytes to be activated by locally expressed signaling molecules. Leukocyte activation increases the avidity of integrins for their immunoglobulin-like counter-receptors on the endothelium, thereby strengthening adhesion. Several general reviews on the selectins have appeared [1–4]. This paper will focus on recent studies of P-selectin.

Each of the selectins contains an *N*-terminal carbohydrate-recognition domain like those in Ca^{2+}-dependent animal lectins, followed by an epidermal growth factor-like motif, a series of consensus repeats related to those in complement-binding proteins, a transmembrane domain, and a short cytoplasmic tail. The selectins mediate cell adhesion through Ca^{2+}-dependent interactions of the carbohydrate-recognition domain with cell surface oligosaccharides. The nature of the carbohydrate ligands for the selectins has elicited intense interest. All three selectins bind with low affinity to sialylated, fucosylated tetrasaccharides such as sialyl Lewis x (sLe^x; Siaα2-3Galβ1–4[Fucα1–3]GlcNAc-R). Both the sialic acid and fucose moieties are required for recognition. Because sLe^x is found on many glycolipids and glycoproteins of myeloid cells, a frequent interpretation of early studies was that this tetrasaccharide represented the physiological ligand for E- and P-selectin. However, accumulating data from several laboratories indicate that the selectins bind to specific glycoprotein ligands with affinities that are much higher than those to simple oligosaccharides such as sLe^x.

P-selectin isolated from human platelet membranes binds to a limited number of sites on human neutrophils and HL-60 cells; these sites are protease-sensitive, suggesting that they are glycoproteins rather than glycolipids [5]. Recombinant forms of P-selectin lacking the transmembrane domain are monomeric, whereas the membrane protein is oligomeric in detergent levels below the critical micellar concentration [6]. In studies of equilibrium binding to myeloid cells, both the monomers and oligomers appear to interact with the same sites, as assessed by competitive binding analysis. The monomers bind to 25,000 sites per neutrophil with an apparent K_d of only 50–100 nM [6]. This is a remarkably high affinity for either an adhesion molecule or a monomeric lectin, suggesting that the ligands on intact myeloid cells identified by this assay have structures more complex than sLe^x. Consistent with this interpretation, Chinese hamster ovary cells transfected with a fucosyltransferase display only low affinity binding sites for fluid-phase P-selectin, even though they express higher levels of cell surface sLe^x than do myeloid cells [7].

We have identified a high affinity glycoprotein ligand for P-selectin in extracts of neutrophils and HL-60 cells [8,9]. The ligand, a minor component of total plasma membrane glycoproteins, is an extensively sialylated dimer consisting of two disulfide-linked subunits of M_r 120,000. It carries α2–3-linked sialic acids and the sLe^x antigen. Almost all of the sialic acids are unmodified N-acetylneuraminic acid. The protein contains a limited number of N-linked glycans that are not required for binding to P-selectin. In contrast, it contains a large number of sialylated, O-linked oligosaccharides that are released by β-elimination. The ligand is cleaved by the enzyme O-sialoglycoprotease from *Pasteurella hemolytica*, which recognizes only proteins containing clustered, sialylated O-linked glycans. Treatment of HL-60 cells with this enzyme eliminates the high affinity binding sites for P-selectin and abolishes adhesion to immobilized P-selectin, without affecting total surface expression of sLe^x. These data suggest that the ligand accounts for the high affinity binding sites on intact myeloid cells, and may play an important role in mediating cell adhesion to P-selectin. Recently, a cDNA encoding the protein component of the ligand was isolated by expression cloning in cells cotransfected with a fucosyltransferase [10]. The ligand, termed P-selectin Glycoprotein Ligand 1 (PSGL-1), is a type I membrane protein that includes 15 decameric repeats rich in threonine residues; the repeats have many potential sites for addition of clustered O-linked oligosaccharides that would render the protein sensitive to O-sialoglycoprotease. The expressed protein has both N- and O-linked carbohydrate and binds to myeloid cells.

How closely the carbohydrate modifications of the recombinant protein match those of the native protein in myeloid cells is unknown.

The structural basis for high affinity binding of PSGL-1 to P-selectin remains to be determined. The *O*-linked oligosaccharides may contain features not found on sLe^x or related known sialylated, fucosylated lactosaminoglycans. Closely packed glycans might also form a rigid "clustered saccharide patch" where components of multiple oligosaccharides create a unique recognition structure [9]. Structural characterization of the individual oligosaccharides and determination of the sites of attachment will help explain why this ligand interacts so well with P-selectin.

High affinity glycoprotein ligands for both L- and E-selectin have also been identified [11-14]. At least some of these ligands also contain numerous sialylated, *O*-linked oligosaccharides. Although the glycoprotein ligands are attractive candidates for mediating selectin-dependent cell adhesion, further studies are required to confirm this hypothesis. It remains possible that larger numbers of low affinity interactions also contribute to cell adhesion. The mechanisms by which selectins mediate cell adhesion under shear forces are even less well understood. It is likely that high on- and off-rates of bond formation are required, but the structural features that favor such kinetics are unknown. Many factors such as receptor and ligand length, mobility, and clustering may contribute to bond formation, and hence adhesion, under shear conditions. P-selectin itself is a rigid, asymmetric protein that should project the carbohydrate-recognition domain approximately 40 nm from the cell surface where it might rapidly contact a ligand on a circulating leukocyte [6].

The regulated expression of the selectins plays a critical role in controlling the duration of leukocyte adhesion to the blood vessel wall. The control of expression of P-selectin is particularly complex. The protein is constitutively synthesized by megakaryocytes and venular endothelial cells, where it is delivered to secretory granules (α granules in platelets and Weibel-Palade bodies in endothelial cells) by a sorting signal in the cytoplasmic domain [15]. Upon stimulation of these cells by agonists such as thrombin or histamine, P-selectin is redistributed to the cell surface as granule membranes fuse with the plasma membrane. On activated endothelial cells, P-selectin is then rapidly internalized, a mechanism to limit the adhesive phenotype of the vessel wall [16]. Internalization occurs in clathrin-coated pits (H. Setiadi and R. P. McEver, unpublished observations) and requires a signal in the cytoplasmic domain ([17] and H. Setiadi and R.P. McEver, unpublished observations). Although antibody binding studies have suggested that P-selectin is efficiently recycled into Weibel-Palade bodies following endocytosis [17], P-selectin antigen disappears in dermal venules following intradermal injection of agents that induce degranulation, suggesting rapid degradation following endocytosis [18,19]. The latter observations may reflect a signal in the cytoplasmic domain that mediates rapid movement of P-selectin from endosomes to lysosomes (Green SA, Setiadi H, McEver RP, and Kelly RB. J. Cell Biol. 1994;124:435–448). Rapid delivery of P-selectin from endosomes to lysosomes is an effective means for limiting recycling to the plasma membrane, and hence, for controlling the duration that the endothelium remains adhesive for leukocytes. Thus, the movements of P-selectin within cells are controlled by information in the cytoplasmic domain that directs sorting into secretory granules, endocytosis, and targeting to lysosomes. The structural features in the cytoplasmic domain required for each of these signals, and the degree to which the signals overlap, are unknown. The single cysteine in the cytoplasmic domain is acylated [20]. The cytoplasmic domain is also phosphorylated, although the sites of phosphorylation are controversial [21,22]. Whether these posttranslational modifications affect the trafficking of P-selectin is unclear.

Inflammatory cytokines increase synthesis of P-selectin in at least some tissues, providing an additional level of regulation [23-25]. Transcripts in both platelets and endothelial cells have been identified that encode an alternatively spliced soluble form of P-selectin lacking the transmembrane domain [26,27]. The physiologic significance of soluble P-selectin is unknown. Because the sorting signal no longer faces the cytoplasm, this protein is constitutively secreted rather than sorted into secretory granules [15]. Very low levels of P-selectin antigen are found in human plasma, and the levels may increase slightly in some thrombotic or inflammatory disorders [6,28-30].

The 5′ flanking region of the human P-selectin gene has been cloned and sequenced [31]. Transcription of the gene is initiated from multiple sites, consistent with the lack of a canonical TATA box that would facilitate transcription initiation at a single site. A relatively short segment of the 5′ flanking region directs specific expression of a reporter gene in cultured endothelial cells, but not in several other cells tested, suggesting that it contains at least some of the elements required for cell type-specific expression. The region contains a number of candidate regulatory elements, including a GATA element that was demonstrated to be functional. A consensus site for binding of members of the NFκB/rel family of transcription factors is also present; whether this site mediates the cytokine-induced increase in P-selectin synthesis observed in some tissues is unknown [23-25].

Dysregulated expression of selectins is a likely mechanism for contributing to the pathogenesis of some inflammatory and thrombotic disorders. In vitro, oxygen radicals cause P-selectin to remain on the surface of endothelial cells for several hours, probably by impairing endocytosis [32]. In vivo, monoclonal antibodies to P-selectin reduce tissue injury in animal models of lung damage [33] and ischemia-reperfusion injury [34,35]. Prolonged expression of P-selectin on the endothelial cell surface at the sites of tissue damage has been documented, consistent with a role for the protein in mediating pathologic leukocyte adhesion. The antibodies presumably inhibit the initial rolling of neutrophils on venules where P-selectin has been inappropriately expressed. Indeed, other studies confirm a role for P-selectin in mediating leukocyte rolling on venules of tissues subjected to mild injury following exteriorization [36-38]. Additional in vivo studies, coupled with further biochemical analysis, should enhance our understanding of the role of P-selectin and the other selectins in physiologic inflammation and hemostasis, and suggest approaches to inhibit selectin function during pathologic inflammation, thrombosis, and tumor metastasis.

Acknowledgments

Work in the author's laboratory was supported by National Institutes of Health grants HL 34363 and HL 45510.

References

1. McEver RP. Curr Opin Cell Biol 1992;4:840-849.
2. Lasky LA. Science 1992;258:964-969.
3. Bevilacqua MP, Nelson RM. J Clin Invest 1993;91:379-387.
4. McEver RP. Selectins. Curr Opin Immunol 1994;6:75–84.
5. Moore KL, Varki A, McEver RP. J Cell Biol 1991;112:491-499.
6. Ushiyama S, Laue TM, Moore KL, Erickson HP, McEver RP. J Biol Chem 1993;268:15229-15237.
7. Zhou Q, Moore KL, Smith DF, Varki A, McEver RP, Cummings RD. J Cell Biol 1991;115:557-564.
8. Moore KL, Stults NL, Diaz S, Smith DL, Cummings RC, Varki A, McEver RP. J Cell Biol 1992;118:445-456.
9. Norgard KE, Moore KL, Diaz S, Stults NL, Ushiyama S, McEver RP, Cummings RD, Varki A. J Biol Chem 1993;268:12764-12774.
10. Sako D, Chang X-J, Barone KM, Vachino G, White HM, Shaw G, Veldman GM, Bean KM, Ahern TJ, Furie B, Cumming DA, Larsen GR. Expression cloning of a functional glycoprotein ligand for P-selectin. Cell 1993;75:1179–1186.
11. Imai Y, Singer MS, Fennie C, Lasky LA, Rosen SD. J Cell Biol 1991;113:1213-1222.
12. Lasky LA, Singer MS, Dowbenko D, Imai Y, Henzel WJ, Grimley C, Fennie C, Gillett N, Watson SR, Rosen SD. Cell 1992;69:927-938.
13. Levinovitz A, Mühlhoff J, Isenmann S, Vestweber D. J Cell Biol 1993;121:449-459.
14. Baumhueter S, Singer MS, Henzel W, Hemmerich S, Renz M, Rosen SD, Lasky LA. Science 1993;262:436-438.
15. Disdier M, Morrissey JH, Fugate RD, Bainton DF, McEver RP. Mol Biol Cell 1992;3:309-321.
16. Hattori R, Hamilton KK, Fugate RD, McEver RP, Sims PJ. J Biol Chem 1989;264:7768-7771.
17. Subramaniam M, Koedam JA, Wagner DD. Mol Biol Cell 1993;4:791-801.
18. Smith CH, Barker JNWN, Morris RW, MacDonald DM, Lee TH. J Immunol 1993;151:3274-3282.
19. Silber A, Newman W, Reimann KA, Hendricks E, Walsh D, Ringler DJ. Lab Invest 1994;70:163–175.
20. Fujimoto T, Stroud E, Whatley RE, Prescott SM, Muszbek L, Laposata M, McEver RP. J Biol Chem 1993;268:11394-11400.
21. Fujimoto T, McEver RP. Blood 1993;82:1758-1766.
22. Crovello CS, Furie BC, Furie B. J Biol Chem 1993;268:14590-14593.
23. Weller A, Isenmann S, Vestweber D. J Biol Chem 1992;267:15176-15183.
24. Sanders WE, Wilson RW, Ballantyne CM, Beaudet AL. Blood 1992;80:795-800.
25. Hahne M, Jäger U, Isenmann S, Hallmann R, Vestweber D. J Cell Biol 1993;121:655-664.
26. Johnston GI, Cook RG, McEver RP. Cell 1989;56:1033-1044.
27. Johnston GI, Bliss GA, Newman PJ, McEver RP. J Biol Chem 1990;265:21381-21385.
28. Dunlop LC, Skinner MP, Bendall LJ, Favaloro EJ, Castaldi PA, Gorman JJ, Gamble JR, Vadas MA, Berndt MC. J Exp Med 1992;175:1147-1150.
29. Katayama M, Handa M, Araki Y, Ambo H, Kawai Y, Watanabe K, Ikeda Y. Br J Haematol 1993;84:702-710.
30. Wu G, Li F, Li P, Ruan C. Haemostasis 1993;23:121-128.
31. Pan J, McEver RP. J Biol Chem 1993;268:22600-22608.

32. Patel KD, Zimmerman GA, Prescott SM, McEver RP, McIntyre TM. J Cell Biol 1991;112:749-759.
33. Mulligan MS, Polley MJ, Bayer RJ, Nunn MF, Paulson JC, Ward PA. J Clin Invest 1992;90:1600–1607.
34. Weyrich AS, Ma X, Lefer DJ, Albertine KH, Lefer AM. J Clin Invest 1993;91:2620-2629.
35. Winn RK, Liggitt D, Vedder NB, Paulson JC, Harlan JM. J Clin Invest 1993;92:2042-2047.
36. Bienvenu K, Granger DN. Am J Physiol Heart Circ Physiol 1993;264:H1504-H1508.
37. Dore M, Korthuis RJ, Granger DN, Entman ML, Smith CW. Blood 1993;82:1308-1316.
38. Mayadas TN, Johnson RC, Rayburn H, Hynes RO, Wagner DD. Cell 1993;74:541-554.

Endothelium-Derived Factors and Vascular Functions. T. Masaki, ed.   233

Selectin-sialomucin interactions

Laurence A. Lasky
Department of Immunology, Genentech, Inc., 460 Pt. San Bruno Blvd., San Francisco, CA 94080, U.S.A.

Summary

The presentation of carbohydrate side chains to the selectins by the sialomucin adhesion family demonstrates a new concept in cell adhesion. The reason for this unique type of cell adhesion mechanism probably lies in the fact that the capture of leukocytes by the endothelium adjacent to an inflammatory site is a novel situation in cell biology. Thus, this is one of the few cases where a rapidly speeding cell must be efficiently attached to an endothelial site in a timely manner. The potentially rapid on rate of selectin-carbohydrate binding, together with the clustering of such carbohydrates in a novel manner by the sialomucins, could serve to overcome both the velocity and avidity issues of such an important interaction. While it may be difficult to experimentally distinguish between the two proposed models of carbohydrate presentation by the sialomucins, such an understanding may ultimately allow for the production of higher affinity inhibitors of selectin mediated cell adhesion and inflammation.

Introduction

The migration of leukocytes to regions of normal and abnormal inflammation involves a complex interplay between a diversity of adhesion and signaling pathways [1,2]. The combinatorial matrix formed by this plethora of mediators in part determines which leukocyte subset invades which tissue and at what time. This allows for an acute response mediated by the neutrophilic subset of leukocytes to be rapidly directed against an invading organism, while a leukocyte response mediated by macrophage/monocytes and T and B cells can occur at a later time. This temporal control permits the organism to mount both rapid, non-specific defenses as well as long-lived, antigen specific defenses, thus insuring both early protection as well as protection against a pathogen that may be encountered again. It is also likely that similar adhesion and signaling cascades are involved with the influx of other leukocyte subtypes to different areas of acute or chronic inflammation. Thus, for example, it is likely that a different combination of adhesion and signaling molecules induces the influx of eosinophils to an allergic site, such as the lung, than those which induce the migration of neutrophils to a lung which is undergoing an acute, pathogen-mediated immune response. It is possible, therefore, that an understanding of the various molecules and physiological mechanisms used during different inflammatory responses may provide keys to the production of specific anti-inflammatory compounds which inhibit some types of inflammation but not others.

The initial contact between the leukocyte and the endothelium adjacent to an

234

inflammatory site manifests itself in the physiological response known as leukocyte rolling [3,4]. Intravital microscopy studies have shown that leukocytes are decelerated approximately two orders of magnitude by relatively low affinity adhesive interactions between the white cells and the endothelium such that the leukocytes appear to "roll" along the vascular wall. A number of laboratories have demonstrated that these lower affinity adhesive interactions are mediated by a group of adhesion molecules that have been termed the selectins. The selectins are lectin-containing adhesion molecules [5–8] that mediate leukocyte rolling by the recognition of sialylated, fucosylated and, in some cases, sulfated carbohydrates that are presented in a tissue or cell specific manner on the leukocyte or endothelial cell surface [9–12]. The specific recognition of such carbohydrates is the critical initiating step of the inflammatory cascade, and the subject of this review will be the mechanisms of carbohydrate ligand recognition by the selectins.

Detailed aspects of selectin-carbohydrate recognition: molecular modeling and mutagenesis

Initial cDNA cloning studies revealed that the selectins contained a type C, or calcium dependent, lectin (carbohydrate binding) domain at their N-termini, and subsequent work revealed that this lectin domain was the major mediator of cell adhesion induced by selectins [5–8]. The minimal carbohydrate epitope recognized by all three selectins was the sialyl Lewis X (sLe^x) type compound [9–11]. This carbohydrate was minimally defined as a tetrasaccharide sugar containing the following structure: sialic acid alpha 2–3 galactose beta 1–4 (fucose alpha 1–3) N acetyl glucosamine, with both the sialic acid and fucose residues being critical for selectin recognition. Extensive modeling and mutagenesis studies of the E and P selectin lectin domains revealed a face of both proteins that appeared to be involved with ligand recognition [13,14]. From the dimensions of this hypothetical ligand recognition domain, it appeared likely that only one carbohydrate was bound per lectin motif, and Fig. 1 shows a model of how the sLe^x tetrasaccharide could dock with the E selectin lectin domain. This figure demonstrates three interesting principles. The first is that the lectin appears to recognize the carbohydrate through a surface phenomenon, with no obvious pockets being involved with recognition. This result might be expected for interactions which occur with a very rapid on rate. The second is that the fucose of the tetrasaccharide appears to interact with the calcium loop, a result that is consistent with the co-crystal structure of a related type C lectin, the mannose binding protein complexed with mannose, and which may explain the calcium dependence of selectin recognition. Finally, the results suggest that the avidity of the interaction between the selectins and their cognate carbo-hydrate ligands may be quite low when the carbohydrate is presented as a soluble molecule, perhaps due to the surface-type of association hypothesized above. This latter conjecture is supported by a large number of in vitro binding studies which

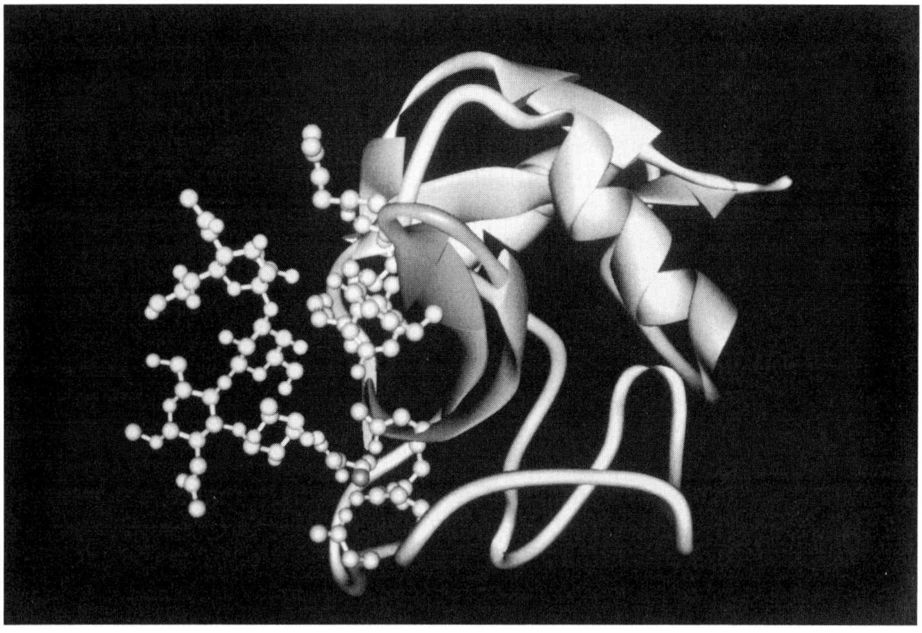

Fig. 1. A model of the lectin domain of E selectin with an sLe^X carbohydrate monomer positioned as it might be during the approach of a leukocyte to the endothelium. Note the highlighted amino acid side chains that have been found by mutagenesis to affect carbohydrate recognition and cell adhesion [13,14]. In addition, it should be noted that a relatively simple surface interaction is hypothesized to take place when sLe^X binds to the selectin.

demonstrate that the sLe^x carbohydrate reacts with very low affinity in assays using either recombinant selectin-IgG chimeras or cell-based adhesion assays.

Sialomucin-like carbohydrate presentation scaffolds

The structure-function and modeling studies of E and P selectin presented a quandary. The apparent surface interaction between sLe^x-related carbohydrates and the lectin domain was consistent with relatively weak adhesion, and the low avidities that were seen in in vitro binding assays which utilized sLe^x-related tetrasaccharide as inhibitors of selectin-mediated adhesion were consistent with this notion [15,16,17]. These data were in stark contrast with the apparent biology of selectin-mediated adhesion in vivo, where a leukocyte that was streaking past an inflammatory site was effectively caught by adhesive interactions between the selectins and the carbohydrates on the cell surface. While it seemed clear that the rapid recognition of the carbohydrate ligand by the selectin could be accomplished by the interaction of the sugar with the surface of the selectin lectin motif, the required strong avidity of this interaction in the presence of the high shear forces

236

induced by vascular flow seemed to mitigate against a role for this type of apparently weak binding interaction.

One mechanism for the enhancement of avidity between two interacting compounds would be to cluster the components in some specific manner. It was thus possible that the avidity of the selectins for their carbohydrate ligands might be greatly enhanced by clustering of these two types of molecules. While data are currently not available for E or P selectin, L-selectin may in fact require clustering for its function. Immunoelectron micrographs of L-selectin on neutrophil surfaces has demonstrated that this selectin is morphologically clustered on the tips of microvillar processes on the cell surface [18]. In addition, removal of the L-selectin cytoplasmic tail, which could function in clustering of the protein by interactions with the cytoskeleton, resulted in mutant proteins that were unable to adhere to endothelial cells in vitro or in vivo [19].

These data suggested that selectins may recognize a clustered array of carbohydrate ligands, and molecular characterization of three endothelial ligands for L-selectin and one leukocyte ligand for P selectin has recently shown this to be the case. The first of the L-selectin endothelial ligands to be described is a sulfated glycoprotein of 50 kD molecular weight termed Sgp50 or GlyCAM 1. cDNA cloning of GlyCAM 1 has shown that this L-selectin ligand is in fact an apparently secreted mucin-like molecule which contains a large number of clustered, O-linked carbohydrate side chains that act as L-selectin ligands [20,21]. These side chains appear to be clustered in two, relatively well conserved domains [22]. A second L-selectin ligand, originally termed Sgp90, has been characterized as the vascular form of CD34 [23]. Like GlyCAM 1, CD34 is a mucin-like cell surface molecule that contains a relatively extended O-linked carbohydrate-rich domain at its N-terminus. A third type of L-selectin ligand appears to be the immunoglobulin-like adhesion molecule, MadCAM ([24], M. Briskin-pers. comm.). This molecule is a chimera which contains both immunoglobulin domains that are recognized by the $\alpha_4\beta_7$ and $\alpha_4\beta_1$ [25] integrin s and a mucin-like domain that appears to present clustered O-linked carbohydrates to the L-selectin lectin domain. Finally, neutrophils contain a ~120 kD cell surface glycoprotein, termed the P Selectin Glycoprotein Ligand (PSGL), which is also a mucin-like molecule that appears to present clustered O-linked carbohydrates to P and possibly E selectin ([26–28] T.Ahern, pers. comm.). These data are summarized in Fig. 2, and because all of these mucin-like scaffolds appear to present sialylated carbohydrate side chains to the selectins, I will refer to them here as the sialomucin-like family of adhesion molecules.

Models of selectin-sialomucin interactions

The current evidence for the involvement of specific presentation of carbohydrate side chains to the selectins by the sialomucin adhesion family revolves around three findings. In the first, a discrepancy has been found between the number of peripheral lymph node (PLN) high endothelial venule (HEV) proteins that appear to contain a specific carbohydrate epitope and the number of PLN HEV proteins

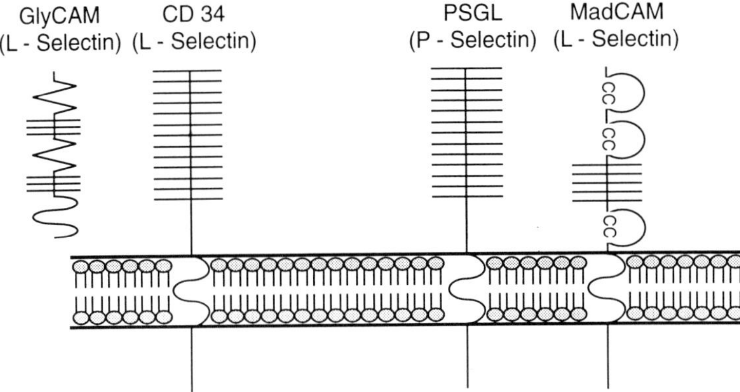

Fig. 2. The four members of the sialomucin-like adhesion family [21,22,23,24, T. Ahern-personnel communication]. The mucin-like molecules that present carbohydrate ligands to the selectins are shown in a hypothesized extended bottle brush like configuration with a multivalent presentation of *O*-linked carbohydrates. GlyCAM 1 is the only member of the family currently known to not have an obvious transmembrane binding domain, and it may function as a soluble molecule. The loop-like structures in MadCAM are immunoglobulin-like domains that are disulfide bonded.

that can be recognized by a recombinant form of L-selectin [29]. The monoclonal antibody MECA 79 appears to recognize a sulfated carbohydrate epitope that is found on a number of glycoproteins expressed on HEV of PLN [20]. Because MECA 79 blocks adhesion of lymphocytes to PLN HEV, it is likely that this carbohydrate is a ligand for L-selectin-mediated adhesion [30]. Of this diversity of MECA 79$^+$ glycoproteins, only three, GlyCAM 1, CD34 and an unknown molecule of >200 kD molecular weight, appear to interact in solution with L-selectin [20]. One interpretation of these data is that while many proteins contain the sulfated carbohydrate epitope recognized by L-selectin, only a subset of them present this ligand to the L-selectin lectin domain in an appropriate manner. In a second example, it has been shown that, while many myeloid proteins contain the sLe^x carbohydrate epitope that is recognized by P selectin, only one protein of ~120 kD molecular weight appears to react with immobilized P selectin [28]. In addition, treatment of neutrophils or HL60 cells with a protease, *O*-sialoglycoprotease, that specifically degrades mucins results in a loss of P selectin binding and a disappearance of the ~120 kD protein. Interestingly, cells treated with this enzyme still contain the vast bulk of the sLe^x carbohydrate epitope found on untreated cells, suggesting that the mere presence of the carbohydrate is insufficient for P selectin binding and that only a small fraction of sLe^x is on this apparently mucin-like molecule. Finally, expression of the cloned PSGL mucin-like glycoprotein (see above) in cells expressing the sLe^x carbohydrate results in P selectin binding, while expression of another sialomucin, CD43 or leukosialin, in these sLe^x expressing cells does not result in P selectin-mediated adhesion (T. Ahern, pers. comm.). Together, these results suggest that only the correct mucin-like glycoprotein can appropriately present carbohydrate ligands to the selectins.

How might these sialomucin-like adhesion molecules present carbohydrate side chains to the selectins? From the previously described modeling and mutagenesis studies, it seems likely that a specific configuration of carbohydrates is presented to a relatively confined and unique site on the surface of the selectin lectin domain. When taken together with the data on the requirement for presentation of the carbohydrates by the sialomucins, this specific carbohydrate configuration could occur via a number of different mechanisms. Here I will present two models, the homomeric and the heteromeric clustering models, which attempt to bring together the apparent requirements for specific configuration or clustering and presentation of carbohydrate ligands to the selectins. The major conceptual difference between the homomeric and heteromeric clustering models of carbohydrate presentation lies in the question of whether each *O*-linked side chain of the sialomucin can function as an independent ligand (homomeric clustering) for the selectins or if several adjacent *O*-linked structures form a conformationally unique patch that results in a novel carbohydrate configuration that would not be formed by each single monomeric side chain (heteromeric clustering).

In the case of homomeric clustering, the rigid, extended structure formed by the mucin-like domain would present a number of independently functional *O*-linked carbohydrate side chains in a unique, clustered configuration that would presumably have increased avidity for a similarly clustered array of selectins on the opposing cell surface. The available data suggest that a high density of *O*-linked chains alone is insufficient for effective recognition by at least P and E selectin, since neither selectin binds to the sialomucin leukosialin (CD43) when it is found normally expressed on leukocytes or when it is transiently expressed in mammalian cells, in spite of the fact that this mucin, as well as a number of other non-mucin glyco-proteins, contain the selectin ligand, sLe^x. Therefore, a unique configuration of *O*-linked side chains would have to be formed by the correct sialomucin to allow for a high affinity interaction. For example, the appropriate sialomucin may present the independently functional carbohydrate ligands to the selectins in a specific geometric configuration that could effectively bind to a similarly configured cluster of selectins (for one such configuration, see Fig. 3, left panel). Such a "lock and key" type of mechanism would allow for a higher avidity interaction through a specific fit between the cluster of selectins on one cell and the cluster of *O*-linked side chains on the other [31].

On the other hand, heteromeric clustering would form the appropriate ligand configuration via a different geometric mechanism (previously proposed in reference [28]). In this case, the required side chains, including for example the sialic acid, fucose, and, in the case of L-selectin, the sulfate, would be found on adjacent or nearby *O*-linked structures. The function of the mucin chain would be to configure these closely linked non-functional side chains into a configuration that would become an effective selectin ligand (Fig. 3B). It is likely that this mechanism would allow for the production of sugar chain configurations not found in oligosaccharides, and these unique configurations might result in higher affinity "patches" that would be recognized by the lectin domain. The mucin could also present a number of these uniquely configured patches such that it could combine the unique structure

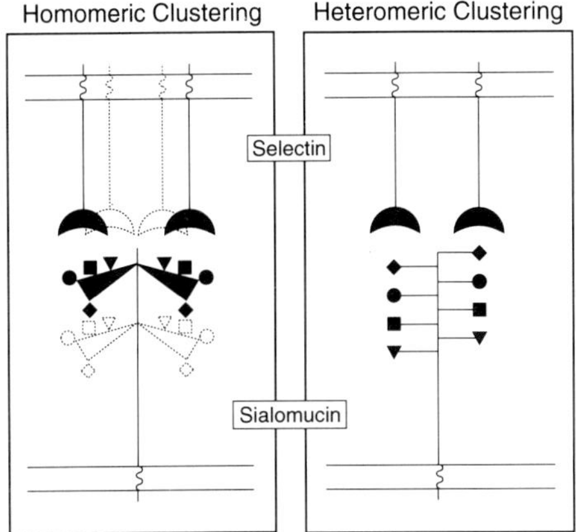

Fig. 3. Models of carbohydrate presentation by the sialomucin-like adhesion molecules. Left: Homomeric clustering where each carbohydrate side chain is an independent ligand for selectin binding which is positioned by the mucin in a specific manner. Shown here is a trivalent positioning that could bind to similarly oligomerized selectins on the opposing cell. Right: Heteromeric clustering where a configuration of carbohydrates is uniquely constructed by the sialomucin-like scaffold [28]. Here is illustrated a unique tetrasaccharide produced by four adjacent *O*-linked carbohydrates which has a structure that is not possible in a normally-linked carbohydrate.

produced by the adjacent side chains in a clustered array. This would allow for a further enhancement of avidity, similar to that proposed for homomeric clustering.

References

1. Butcher EC. Adv Exp Med Biol 1992;323(181):181–194.
2. Lasky LA. Science 1992;258(5084):964–969.
3. Ley K et al. Blood 1991;77(12):2553–2555.
4. von Andrian U, Chambers J, McEvoy L, Bargatze R, Arfors K, Butcher E Proc Natl Acad Sci USA 1991;88:7538–7542.
5. Lasky LA et al. Cell 1989;56(6):1045–1055.
6. Bevilacqua MP et al. Science 1989;243(4895):1160–1165.
7. Johnston GI, Cook RG, McEver RP. Cell 1989;56(6):1033–1044.
8. Tedder TF et al. J Exp Med 1989;170(1):123–133.
9. Phillips ML et al. Science 1990;250(4984):1130–1132.
10. Walz G et al. Science 1990;250(4984):1132–1135.
11. Brandley BK, Swiedler SJ, Robbins PW. Cell 1990;63(5):861–863.
12. Imai Y, Lasky LA, Rosen SD. Nature 1993;361(6412):555–557.
13. Erbe DV et al. J Cell Biol 1992;119(1):215–227.
14. Erbe DV et al. J Cell Biol 1993;120(5):1227–1235.
15. Foxall C et al. J Cell Biol 1992;117(4):895–902.
16. Nelson RM et al. J Clin Invest 1993;91(3):1157–1166.
17. Mulligan M, Paulson J, Frees S, Zheng Z, Lowe J, Ward P. Nature 1993;364:149–151.
18. Picker LJ et al. Cell 1991;66(5):921–933.
19. Kansas GS et al. J Exp Med 1993;177(3):833–838.
20. Imai Y et al. J Cell Biol 1991;113(5):1213–1221.

21. Lasky LA et al. Cell 1992;69(6):927–938.
22. Dowbenko D, Watson S, Lasky LA. J Biol Chem 1993;268(19):14399–14403.
23. Baumhueter S, Singer M, Henzel W, Hemmerich S, Renz M, Rosen S, Lasky LA. Science 1993;262:436–438.
24. Briskin M, McEvoy L, Butcher E. Nature 1993;363:461–464.
25. Berlin C et al. Cell 1993;74(1):185–195.
26. Moore KL, A Varki, McEver RP. J Cell Biol 1991;112(3):491–499.
27. Moore KL et al. J Cell Biol 1992;118(2):445–456.
28. Norgard K, Moore K, Diaz S, Stults N, Ushiyama S, McEver R, Cummings R, Varki, A. J Biol Chem 1993;268(17):12764–12774.
29. Berg EL et al. J Cell Biol 1991;114(2):343–349.
30. Streeter P, Rouse B, Butcher E. J Cell Biol 1988;107:1853–1862.
31. Drickamer K. Biochem Soc Trans 1993;21:456–459.

Endothelium-Derived Factors and Vascular Functions. T. Masaki, ed.

241

An important role of adhesion molecules (LFA-1 and ICAM-1) in endothelial cell injury caused by activated neutrophils

Sei-itsu Murota, Hiroshi Fujita, Ikuo Morita and Yoshiyuki Wakabayashi

Department of Physiological Chemistry, Graduate School, Tokyo Medical and Dental University, Yushima, Bunkyo-ku, Tokyo, Japan

Summary

Treatment of leukocytes with phorbol myristate acetate (PMA) caused significant increases in the expression of adhesion molecules, CD11a, CD11b, CD11c and CD18. The addition of the PMA-stimulated leukocytes to an endothelial cell monolayer caused a significant increase in the intracellular peroxide level of the endothelial cells after 15 minutes and endothelial cell injury after 5 hours. Both the early increase in the peroxide production and the late cell lysis were abolished in the presence of specific antibodies against CD11a, CD11b, CD18 and ICAM-1, but not CD11c. These antibodies affected neither the production of active oxygen species by the leukocytes nor the rate of adhesion of leukocytes to endothelial cells. These data indicate that the adhesion through CD11/CD18-ICAM-1 is necessary for the leukocytes to induce endothelial cell injury as well as peroxide production in endothelial cells. Both the early and late events were only partially blocked by catalase but almost completely abolished by deferoxamine, a chelator of ferrous ions, suggesting that hydroxyl radicals produced by endothelial cells by xanthine oxidase may injure themselves. Pretreatment of endothelial cells with allopurinol, a specific inhibitor of xanthine oxidase, caused significant inhibition of both the early and the late events, suggesting that the binding of adhesion molecules such as ICAM-1 and CD11/CD18, but not CD11c, may trigger the activation of xanthine oxidase of endothelial cells and have the cells to produce more hydrogen peroxide and ferrous ions, followed by producing more hydrogen peroxide. The hydrogen peroxide produced by endothelial cells themselves and by leukocytes may be converted to hydroxyl radicals by ferrous ions, which may cause lethal cell damage. Examination of xanthine oxidase activity in endothelial cells showed that the enzyme activity increased double within 15 minutes after the addition of PMA activated leukocytes.

Introduction

Endothelial cells are multi-functional cells producing many kinds of bioactive substances. Therefore, injury of endothelial cells may easily bring about various vascular disorders.

It has been well known that leukocytes play some important roles in ischemia-reperfusion injury of blood vessels. The initial event of the injury is known to be in adhesion of activated leukocytes to endothelial cells. The fact that some radical scavengers can reduce the ischemia-reperfusion injury suggests that active oxygen

242

radicals have something to do with this endothelial cell injury. To find out the mechanism by which leukocytes injure endothelial cells, we established some in vitro assay systems capable of measuring endothelial cell injury due to the activated leukocytes.

Materials and Methods

Leukocytes were isolated from healthy volunteers. The neutrophil fraction was then purified by using Percoll. The purified leukocytes were then stimulated with PMA, a kind of phorbol esters which caused activation and aggregation of the leukocytes and resulted in having them release active oxygens. The amount of the released active oxygens was measured by luminol dependent chemiluminescence.

Leukocytes were placed on the monolayer of cultured endothelial cells, followed by stimulating them with PMA, and induced endothelial cell injury was examined. For this purpose the endothelial cells were labeled with Cr^{51} beforehand. The radioactivity released out of the injured cells was counted, and we made an index of the injury.

Results and Discussion

The PMA-stimulated leukocytes caused endothelial cell injury, and it was dependent on the dose of PMA added. There was a good correlation between the active oxygen releasing activity of the leukocytes and their cytotoxicity [1]. When leukocytes were placed directly on a monolayer of endothelial cells, and stimulated with PMA (10 ng/ml), a severe endothelial cell injury was observed after 5 hours. Leukocytes alone or PMA alone had no such activity. To confirm that the cytotoxicity due to the PMA-activated leukocytes was actually mediated by active oxygen radicals, we used a special device which had been developed for culturing two different types of cells without having them contact each other named "Intercell", i.e., when leukocytes were placed on a filter which was set apart from the monolayer of endothelial cells, and stimulated with PMA, the endothelial cell injury was almost completely abolished. These results strongly suggest that the substance responsible for the cytotoxicity must be some very short lived, labile substance, presumably active oxygen radicals. We examined these phenomena further from the view point of cell adhesion molecules.

A number of cell adhesion molecules are known to be expressed on the surfaces of both leukocytes and endothelial cells. These adhesion molecules are known to be deeply involved in the tight adherence of leukocytes to endothelial cells. We examined the role of adhesion molecules in the cell injury. The process that we are going to discuss deals with the results from cleaving one of the bindings.

First of all we examined what kinds of adhesion molecules were expressed on the surface of leukocytes after the PMA stimulation. One hour after the PMA stimulation significant increases in the expression of CD11a,b,c and CD18 was observed. Then we examined the effects of cleaving the bindings in which these adhesion molecules were involved on the endothelial cell injury due to the activated leukocytes.

The antibodies against CD11a, CD11b and CD18 but not CD11c caused significant inhibition of endothelial cell injury due to the PMA stimulated leukocytes without affecting active oxygen releasing activity of the leukocytes and their adhesion rate to endothelial cells [2]. The inhibitory effects of the anti-CD11a and anti-CD18 antibodies on the endothelial cell injury were dose dependent, respectively. Non-specific IGg had no such activity at all. The inhibitory effect of anti-ICAM-1 antibody on the endothelial cell injury was also dose dependent.

The results obtained so far suggest that even under such condition as the cleaving one particular binding, the leukocytes adhere to endothelial cells to a similar extent to the control level and release active oxygens to a similar extent to the control level. Thus, the appearance seems to be the same with the control culture, but the endothelial cell injuries are very much suppressed by the loss of only one particular binding. Why could this happen?

To solve this puzzle, first we examined whether the endothelial cell injury occurred from outside of the cells or from inside of the cells. In other words, we examined whether active oxygens attack endothelial cell membrane from outside of the cells or from inside of the cells. To know how large the injury from outside is, we added catalase to the culture medium to see what would happen to the endothelial cell injury. Catalase is an enzyme to metabolize hydrogen peroxide into water. The reason why we used catalase is that the major active oxygen species able to attack endothelial cell membrane from outside must be hydrogen peroxide, because the life spans of other kinds of active oxygens, such as superoxide anions, hydroxylradicals and singlet oxygens, if any, are very short and their diffusable distances are also too limited to reach endothelial cell membranes near by. Therefore, we examined the effect of catalase on the endothelial cell injury due to the PMA-stimulated leukocytes. Since catalase is a macromolecule enzyme, it cannot penetrate cell membrane, but can stay just outside of endothelial cells, therefore, by using catalase we can tell the event that occurs only outside of the cells.

The endothelial cell injury was only partially abolished by the catalase, suggesting that the injury from outside is not so large, at most 40%. The remaining 60% of the injury must be caused by something else other than hydrogen peroxide. As we expected SOD was essentially inactive. On the other hand, deferoxamine mesylate abolished the endothelial cell injury almost completely.

Deferoxamine mesylate is one of the chelating agents of metal ion, since it is a low molecular substance it can freely get into the cells, and can block the production of hydroxyl radicals by chelating ferrous ions. Hydroxyl radicals are the most potent active oxygen able to trigger lipid peroxidation chain reaction, followed by destroying cell membrane. Therefore, we assumed that most of the endothelial cell injury due to the activated leukocytes must occur from inside of the cells by hydroxyl radicals produced by endothelial cell themselves. Regarding the source of hydrogen peroxide, an important direct precursor of hydroxyl radical, we are

244

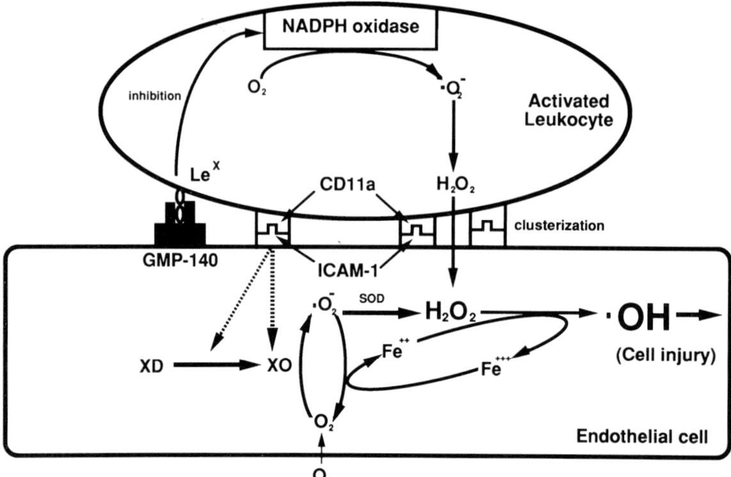

Fig. 1. Mechanisms by which activated leukocytes injure endothelial cells.

assuming two possibilities. One, which is about 40%, is coming from activated leukocytes. The other, which is the remaining 60%, is being produced by endothelial cells themselves by using xanthine oxidase and the pathways shown in Fig. 1. Our working hypothesis was that the binding of these particular adhesion molecules may trigger the activation of xanthine oxidase through some unknown intracellular signaling and have the cells produce more superoxide anions, which causes more production of hydrogen peroxide and more reduction of ferric ions to ferrous ions, which result in more production of hydroxyl radicals by Harber-Weiss reaction (Fig. 1). To examine this hypothesis, next we measured the change in the hydrogen peroxide level in endothelial cells after the addition of PMA-stimulated leukocytes.

To attain this purpose, endothelial cells were prelabeled with DCFDA (Fig. 2). This agent can freely path through cell membrane, however, once the agent gets into the cells, the acetyl groups are immediately hydrolized by esterase in cytosol. The hydrolized agent cannot path through the cell membrane any longer, being trapped within the cells. Then, hydrogen peroxide, if it comes, can immediately activate the hydrolized agent enough to emit fluorescence. Therefore, by measuring the fluorescence emission, we can evaluate the amount of hydrogen peroxide in the cells.

When leukocytes were added to endothelial monolayer which was pre-labeled with DCFDA and activated with PMA, a significant increase in the intracellular fluorescence intensity was observed, after 15 min, suggesting that under this condition the intracellular hydrogen peroxide level increased actually in the endothelial cells. Leukocyte alone as well as PMA alone had no such activity at all.

In the control endothelial cells, 15 min after the addition of PMA stimulated leukocytes, a significant increase in the intracellular fluorescence level was evident. However, in the presence of anti-CD11a antibody, the increase was blocked significantly. The effect of anti-ICAM-1 antibody on the increase in intracellular

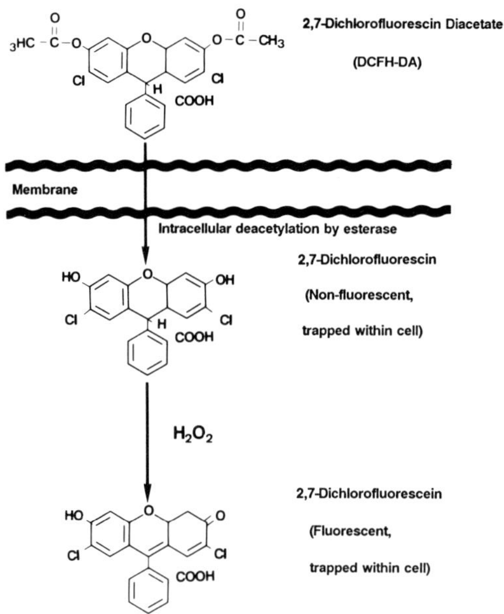

Fig. 2. The method to evaluate intracellular hydrogen peroxide level.

fluorescence intensity was almost similar to that of anti-CD11a antibody.

When we examined the time course of the effects of specific antibodies against various kinds of adhesion molecules on the increase in the intracellular fluorescence level, in the control cultures, after the addition of PMA-activated leukocytes, their intracellular fluorescence levels increased with time up to 15 min. However, in the presence of all of the antibodies against ICAM-1 and CD11/CD18, except anti-CD11c antibody, the increases were significantly abolished (3).

In our assay system, it is evident that the addition of PMA-stimulated leukocytes to an endothelial cell monolayer caused a significant increase in the intracellular hydrogen peroxide level after 15 min, and severe endothelial cell lysis after 5 hours. Both of these early and late events were significantly abolished by specific antibodies against the particular adhesion molecules.

Next we examined the effect of allopurinol, a specific inhibiter of xanthine oxidase, to confirm our working hypothesis. If our working hypothesis is true, allopurinol must block both hydrogen peroxide production and the cell injury.

Allopurinol inhibited the increase in the intracellular fluorescence intensity significantly as we expected. The time course of the allopurinol effect showed that allopurinol inhibited hydrogen peroxide production almost completely at the beginning, but after 10 min the level increased gradually, suggesting that some hydrogen peroxide produced by leukocytes are coming from outside into the cells.

Since allopurinol was found to block hydrogen peroxide production significantly, next we examined whether allopurinol could inhibit endothelial cell injury or not. Allopurinol inhibited the endothelial cell injury significantly, about 60%. Concomi-

246

tant addition of catalase abolished the endothelial cell injury completely, suggesting that our working hypothesis is true.

To confirm our working hypothesis further, we did another experiment in which leukocytes were activated with PMA for 5 hours to have their adhesion molecules expressed on the surface of the membrane, then the leukocytes were fixed with ethanol to kill them. The fixed leukocytes were not able to release active oxygens any longer, but they had some activity to adhere to endothelial cells with their adhesion molecules. Therefore, by using these dead but activated leukocytes we could block the source of hydrogen peroxide from outside completely. If intracellular fluorescence intensity increased even by such dead leukocytes, it means that xanthine oxidase activation was brought about by the binding of these adhesion molecules.

The dead leukocytes were not able to produce active oxygens any longer. However, the endothelial cells to which the dead leukocytes adhered demonstrated hydrogen peroxide production, suggesting that xanthine oxidase was actually activated by the binding of particular adhesion molecules.

To get the exclusive confirmation of our working hypothesis, finally we examined xanthine oxidase activity. The PMA-stimulated leukocytes caused significant increase in xanthine oxidase activity of endothelial cells within 15 min. PMA alone and leukocytes alone had no such activity.

Conclusion

PMA-stimulated leukocytes caused severe injury to endothelial cells. There was a good correlation between the active oxygen releasing activity and the cytotoxicity of the activated leukocytes.

Each antibody against adhesion molecules CD11a, CD11b, CD18, ICAM-1, but not CD11c, inhibited the endothelial cell injury due to the activated leukocytes without affecting the active oxygen releasing activity of the leukocytes and their adhesion rate to endothelial cells.

Most of the endothelial cell injury due to the activated leukocytes occurred from inside of the cells by hydroxyl radicals.

About 60% of the intracellular hydroxyl radicals were produced by the endothelial cells themselves, and the remaining 40% were coming from leukocytes.

There is a possibility that the binding of some particular adhesion molecules, i.e., LFA-1 and ICAM-1 or Mac-1 and ICAM-1, may trigger the activation of xanthine oxydase of the endothelial cells and have the cells produce more active oxygen radicals to injure themselves.

Even killed leukocytes which had been activated with PMA could induce hydrogen peroxide production in endothelial cells.

PMA-activated leukocytes enhanced xanthine oxidase activity in endothelial cells.

References

1. Murota S, Morita I, Suda N. In: Lee KT, Onodera K, Tanaka K (eds) Atherosclerosis II: recent progress in atherosclerosis research. New York: Academy of Sciences, 1990;182–187.
2. Fujita H, Morita I, Murota S. Biochem Biophys Res Commun 1991;177:664–672.
3. Fujita H, Morita I, Murota S. A possible mechanism for vascular endothelial cell injury elicited by activated leukocytes: a significant involvement of adhesion molecules, CD11/CD18 and ICAM-1. Arch Biochem Biophys 1994;309:62–69.

Endothelium-Derived Factors and Vascular Functions. T. Masaki, ed.

Biological significance of L-selectin (LECAM-1) and its ligand

Masayuki Miyasaka[1], Takuya Tamatani[1,2] and Hiroto Kawashima[1]

[1]Department of Immunology, The Tokyo Metropolitan Institute of Medical Science, 3-18-22, Hon-Komagome, Bunkyo, Tokyo 113 and
[2]Japan Tobacco Inc. Pharmaceutical Basic Research Laboratories, 13-2, Fukuura, 1-chome, Kanazawa-ku, Yokohama, Kanagawa 236, Japan

Summary

L-selectin (LECAM-1) is a leukocyte adhesion molecule for endothelium and plays an important role in leukocyte rolling along activated endothelium. We have obtained cDNA encoding rat L-selectin and generated a soluble fusion protein of rat L-selectin and human IgG Fc portion (rLEC-Ig). We then used this protein to characterize the ligands in terms of their biochemistry and tissue distribution. Here we show the heterogeneity of the L-selectin ligands and also discuss the possible significance of L-selectin/ligands interaction.

Introduction

L-selectin (LECAM-1), one of the selectin (LECAM) members, is thought to be the lymphocyte homing receptor that mediates binding of lymphocytes to high endothelial venules of peripheral lymph nodes [1]. Although L-selectin is probably a most well-characterized lymphocyte adhesion molecule, there are still a number of unresolved issues with this molecule, one of which is exact operating mechanism in the lymphocyte-HE cell interaction. This molecule is expressed not only by lymphocytes but also by all other types of leukocytes which in fact never exhibit homing phenomenon under normal conditions.

We have cloned cDNA encoding rat L-selectin and produced a soluble fusion protein of rat L-selectin and human IgG [2], and used it to identify ligand structures recognized by L-selectin [3], and also to produce blocking as well as non-blocking monoclonal antibodies to rat L-selectin [4,5]. By using these tools, we investigated the biological significance of interaction between L-selectin and its ligands.

Results and Discussion

Ligands for L-selectin are present in multiple anatomical locations

The use of soluble form of recombinant adhesion molecules has been proven to be extremely useful in investigation of the distribution of their ligands [6]. Therefore,

250

we generated a soluble recombinant form of L-selectin (rLEC-Ig) in rats and used it as a histochemical probe [3] to examine the systemic localization of ligands for L-selectin in rats. Frozen sections of various tissues were incubated with either rLEC-Ig or a control fusion protein PVR-Ig, and the binding was detected by the use of peroxidase-conjugated anti-human Ig [4]. As reported by Watson et al. [6], HEV in peripheral lymph nodes showed intense staining with rLEC-Ig but not with PVR-Ig.

Interestingly, rLEC-Ig recognized structures in non-lymphoid tissues as well. For instance, the myelin-rich white matter in the central nervous system and the choroid plexus in the brain showed positive staining. While proximal tubules and glomeruli in the renal cortex were completely negative, distal tubules and capillary blood vessels in renal medulla exhibited very intense staining with rLEC-Ig (summarized in Table 1). These results demonstrate that ligands for L-selectin are present in multiple anatomical locations.

Multiple molecular species are present in ligands for L-selectin in peripheral lymph nodes

We [4] and others [6,7] have previously demonstrated that ligands for L-selectin can be readily labeled with sodium [^{35}S]sulfate. As shown in Fig. 1, rLEC-IgG precipitated predominant bands migrating at 50, 95 and 250 kDa, respectively, from [^{35}S]sulfate-labeled lymph node culture supernatant, and 125 and 250 kDa bands from sodium [^{35}S]sulfate-labeled lymph node lysates under reducing conditions, indicating that the 50, 95 and 250 kDa species are readily shed, whereas the 125 kDa species is tightly associated with cells. Since HEV was the only structure that expresses L-selectin ligand in lymph nodes, we interprete these results to mean that the 50, 95 and 250 kDa species are all expressed by HEV and shed from there. Interestingly, reactivity of rLEC-Ig to these multiple molecular species was completely abrogated, when lymph nodes were pretreated with benzyl GalNAc which blocks processing of *O*-linked (mucin-type) sugars, but remained unchanged when the lymph nodes were pretreated with swainsonine which inhibits a *N*-linked sugar processing enzyme α-mannosidase II. These results indicate that all the molecular species detected by rLEC-Ig are modulated with *O*-linked sugars and that rat L-selectin recognizes the *O*-linked sugars but not *N*-linked sugars. Although further study is required, the 50 kDa species most likely represent the rat homologue of GlyCAM-1 [8], whereas the 95 kDa species probably represent rat CD34 [9]. The higher molecular weight species are currently unknown (summarized in Table 2).

Table 1. The presence of L-selectin ligands in various tissues

Tissues	Localization
Lymph nodes	High endothelial venule (HEV)
Brain	White matter, choroid plexus, Purkinje cell
Kidney	Distal tubule, capillaries in medulla

Table 2. Ligands for L-selectin in lymph nodes

Source	Molecular species (kDa)	Sulfation	*O*-linked sugar modification	Ca^{2+}-dependency
LN culture	250 kDa	+	+	+
supernatant	95 kDa	+	+	+
	50 kDa	+	+	+
LN lysates	250 kDa	+	+	+
	125 kDa	+	+	+
	160 kDa	−	+	+

Characterization of L-selectin ligands expressed in the kidney

During the course of histological investigation of the kidney, we found unexpectedly that a staining pattern almost identical to that obtained with rLEC-Ig was obtained with a monoclonal antibody against sulfatides. Both distal tubules and capillary blood vessels in renal medulla were strongly stained with anti-sulfatide antibody, indicating possibly that L-selectin ligands expressed in the kidney are sulfatides. Indeed, our previous study revealed that L-selectin binds to sulfatides [3]. However, whether-or-not the ligands in the kidney actually represent sulfatides needs to be further investigated, since we found recently that the anti-sulfatide monoclonal antibody we used recognizes some sulfated glycoproteins in the kidney (T. Ishimaru, H. Kawashima, Y. Suzuki and M. Miyasaka; unpubl. obs.). The results that HEV was not positively stained with this antibody [4] and that GlyCAM-1 was undetected in the kidney [8] clearly indicate that ligand detected in the kidney is a non-GlyCAM-1 molecule. Biochemical characterization of the ligand in the kidney is in progress.

The presence of ligands for L-selectin may be an important factor that influences the pathogenesis of various inflammatory disorders in the kidney. Although lymphocyte traffic through the kidney is normally very low, it has been reported that induction of chronic damage to the kidney by mechanical obstruction of the ureter results in massive infiltration of mononuclear cells into the renal parenchyma [10]. Collaborative study initiated with a group of nephrologists revealed that drastic changes in the distribution of L-selectin ligands occur subsequently to ligation of the ureter and that mononuclear cell infiltration begins around the distal tubules where ligands for L-selectin is constitutively expressed (K. Shikata, H. Makino, Y. Yuzawa, S. Matsuo and M. Miyasaka; unpubl. obs.). Prominent lymphocyte infiltration is also seen around the distal tubules in various types of interstitial nephritis [11]. These results collectively indicate that the presence of L-selectin ligands may be indeed an important contributing factor in the pathogenesis of leukocyte infiltration into the kidney.

Acknowledgements

This study was supported by grants from Uehara Memorial Foundation, from the Ministry of Health and Welfare, and also by a Grant-in-Aid for Special Project, Cancer Bioscience from the Ministry of Education, Science and Culture.

References

1. Yednock TA and Rosen SD. Adv Immunol 1989;44:313–378.
2. Watanabe T, Song Y, Hirayama Y, Tamatani T, Kuida K, Miyasaka M. Biochim Biophys Acta 1992;1131:321–324.
3. Suzuki Y, Toda Y, Tamatani T, Watanabe T, Suzuki T, Nakao T, Murase, K, Kiso M, Hasegawa A, Tadano-Aritomi K, Ishizuka I, Miyasaka M. Biochem Biophys Res Comm 1993;190:426–434.
4. Tamatani T, Kuida K, Watanabe T, Koike S, Miyasaka M. J Immunol 1993;150:1735–1745.
5. Tamatani T, Kitamura F, Kuida K, Shirao M, Mochizuki M, Suematsu M, Schmid-Schonbein GW, Watanabe K, Tsurufuji S, Miyasaka M. Eur J Immunol 1993;23:2181–2188.
6. Watson SR, Imai Y, Fennie JS, Geoffroy JS, Rosen SD. J Cell Biol 1990;110:2221–2229.
7. Imai Y, Singer MS, Fennie C, Lasky LA, Rosen SD. J Cell Biol 1991;113:1213–1221.
8. Lasky LA, Singer MS, Dowbenko D, Imai Y, Henzel WJ, Grimley C, Fenney C, Gillet N, Watson SR and Rosen SD. Cell 1992;69:927–938.
9. Baumhueter S, Singer MS, Henzel W, Hemmerich S, Renz M, Rosen SD, Lasky LA. Science 1993;262:436–438.
10. Smith JB, McIntosh GH, Morris B. J Pathol 1970;100:21–27.
11. Smythe CM. In: Becker L (ed.) Structural Basis of Renal Disease. New York: Harper & Row, 1968;519.

Endothelium-Derived Factors and Vascular Functions. T. Masaki, ed.

Novel monoclonal antibody (ASH1a/256C) recognizing asialo GM-2 developed on the surface of fatty streak of atherosclerotic aorta

Tatsuya Takano, Masahiro Mori, Keiji Shima and Tsuneo Imanaka

Department of Microbiology and Molecular Pathology, Faculty of Pharmaceutical Sciences, Teikyo University, Sagamiko, Kanagawa 199-01, Japan

Summary

We isolated novel monoclonal antibody which recognizes the fatty streak of atherosclerotic aorta both in human and WHHL rabbits. The following four criteria were adopted to select the hybridoma: (1) Positive against human atherosclerotic aorta; (2) positive against WHHL rabbit atherosclerotic aorta; (3) negative against any kind of macromolecules in serum; and (4) specifically recognizes the surface of fatty streak of both human and WHHL rabbit. This antigenic material must be asialo GM2 and/or asialo GM2 related compounds.

Introduction

We have reported various kinds of monoclonal antibodies which recognized particular lesions of the atheroma, using homogenates of atherosclerotic aorta as immunogen. The first trial in this series of findings was a new monoclonal antibody which recognizes the extracellular regions where lipids deposit in WHHL rabbit [1,2]. This antigenic material was identified as rabbit vitronectin using cDNA cloning technique [3]. Another exciting monoclonal antibody was recognizing peroxidized lipoprotein [4]. The epitope of the antigenic material was peroxidized phosphatidyl choline [5]. In this study, we tried to prepare another novel monoclonal antibody which recognizes the fatty streak of atherosclerotic aorta both in human and WHHL rabbits.

Methods

BALB/c mice were immunized by human atherosclerotic aorta which was provided from Dr. Numano (Tokyo Medical and Dental University). Spleen cells were fused with P3/U1 mouse myeloma cells, and the resulting hybridoma were selected with HAT selection (hypoxanthine aminopterin thymidine) medium. In the first screening, the following three criteria were adopted: (1) hybridoma which was positive against

254

the homogenate of human atherosclerotic aorta by ELISA; (2) hybridoma which was also positive against the homogenate of WHHL rabbit atherosclerotic aorta; and (3) hybridoma should not recognize any kind of macromolecules of serum in both human and rabbits. Second screening was the most important part of the strategy. Strips of arterial wall prepared from WHHL rabbits was blocked by 1% BSA in PBS(-) and incubated with the culture medium of hybridoma and then incubated with ^{125}I-labeled goat anti-mouse Ig(G+M). The auto-radiogram was taken by X-ray film.

Results

To test whether this antibody binds to the fatty streak of the rabbits atherosclerotic aorta in vivo, ^{123}I-labeled monoclonal antibody was introduced into WHHL via ear vein. After 48 hr, the rabbit was killed, the aorta was opened longitudinally and auto-radiogram was taken. This antibody bind to the fatty streak especially in aortic arch, and also in the blanching lesion of the atherosclerotic aorta. This antibody did not bind any portion of normal aorta. This antibody also recognizes the fatty streak of human atherosclerotic aorta in vitro. These results suggested that the antibody recognized the fatty streak of atherosclerotic aorta both in WHHL rabbit and in human.

The antigenic material detected by ELISA was almost eight times higher in atherosclerotic aorta than that in normal aorta. Almost all of the activity was recovered in acetone fraction but not in residual fraction. Recovery of the antigenic material was more than 100% in both normal and atherosclerotic aorta. Latent activity may be developed during the extracting procedure, though the activity could not always be measured quantitatively by ELISA.

It is well known that atherosclerosis developed more severely in aortic arch and

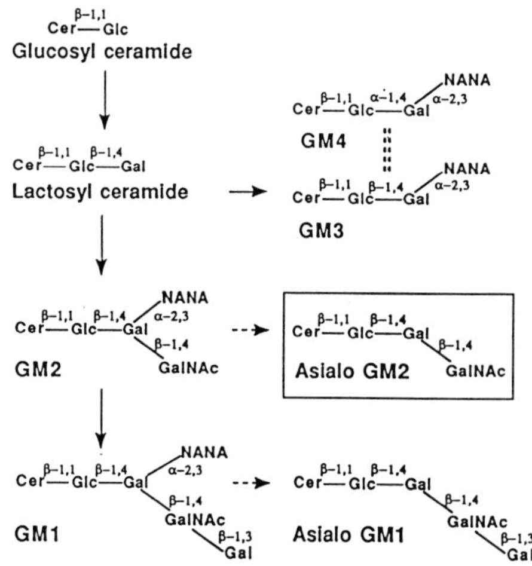

Fig. 1. Comparison of ganglioside structure.

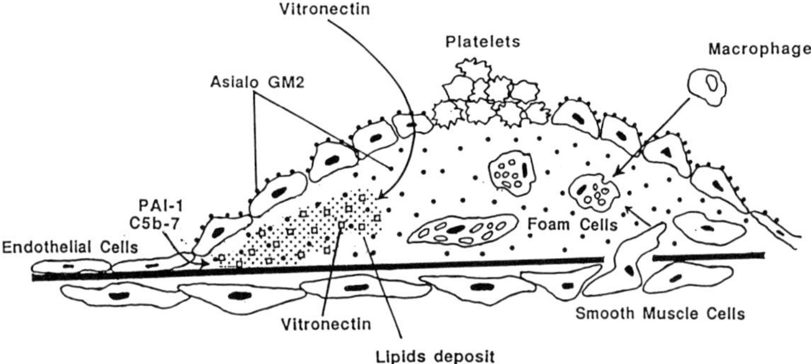

Fig. 2. Asialo GM2 developed on the fatty streak of atherosclerotic aorta. Working hypothesis.

less in abdominal aorta. Arterial wall was cut into 2 cm pieces, and the antigenic activity was measured in each pieces. The activity was high in aortic arch and gradually decreased in the abdominal area as expected.

To examine the epitope site of the antigenic material, competition experiments were carried out. Consequently we found that asialo GM2 (*N*-acetyl-galactosaminyl-b1, 4-galactosyl-b1, 4-glucosyl-b1, 1-ceramide) was a good competitor against this activity. In this condition, however, GM2 and asialo GM1 did not inhibit. Effect of various kinds of gangliosides and proteoglycans on the reactivity against this antibody was summarized in Fig. 1. Competitive inhibitory activity by asialo GM2 was especially high and followed by asialo GM1. The other ganglio family ganglioside, such as GM2, disialoganglioside (GD1a), trisialoganglioside (GT1a, GT1b), ceramide consisting of different chain length of fatty acids, lactosyl ceramide (asialo GM3) were all negative. Globo family gangliosides such as globo triasyl ceramide, globo tetrasyl ceramide had no reactivity was detected. Suggesting that the epitope against this antibody was GalNAc-b1, 4-Gal site of asialo GM2 and/or asialo GM2 related compounds.

To examine that the antigenic material is asialo GM2 itself or asialo GM2 related compounds, acetone extracted fraction was applied on silica TLC. By orcinol reaction, one asialo GM2 and three other minor components were developed. After immuno blotting analysis, the antibody recognized all of these four components, though the reactivity was a little weak. Suggesting that one of the antigenic material must be asialo GM2 and that three other minor components were at least existing.

Discussion

Recently, it was reported that the several gangliosides such as GM1, GM2 and GM3 deposited high in amount in atherosclerotic aorta. It was also known that proteo-glycans such as heparan sulfate and chondroitin sulfate deposited in atherosclerotic lesions. They were all different from antigenic materials recognized by the antibody. If one compares with the analogy of development of asialo gangliosides in macrophages activated by IFN-γ, TNF-α or LPS, endothelial cells on the suface of atheroma may be activated by some stimulation. In this study, we have isolated the monoclonal antibody which recognizes the fatty streak of atherosclerotic aorta, and found that the antigenic material was asialo GM2 and/or asialo GM2 related compounds (Fig. 2). Further investigation was required to clarify the physio-pathological function of asialo GM2 which was developed in fatty streak of atherosclerotic aorta. We hope this kind of antibody will be able to be applied for imaging diagnosis and also for missile chemotherapy in atherosclerosis.

References

1. Kimura J, Nakagami K, Amanuma M, Ohkuma S, Yoshida Y, Takano T. Virchows Arch A 1986;410:159–164.
2. Nakagami K, Shimazaki O, Sato R, Komine Y, Ohkuma S, Takano T. Am J Pathol 1989;135(1): 93–100.
3. Sato R, Komine Y, Imanaka T, Takano T. J Biol Chem 1990;265(34):21232–21236.
4. Mowri H, Ohkuma S, Takano T. Biochim Biophys Acta 1988;963:208–214.
5. Itabe H, Takeshima E, Iwasaki H, Kimiura J, Yoshida Y, Imanaka T, Takano T. J Biol Chem 1994;269 (in press).

Author index

Akishita, M., 93
Ando, J., 103
Ansari, N., 3
Aoyama, T., 21
Azuma, S., 87

Baird, A., 165
Bardhan, S., 199
Berstein, G., 211
Buga, G.M., 13

Canet, E., 47
Cao, W.-H., 87
Chaki, S., 199

Doi, Y., 21
Dzau, V.J., 131

Eguchi, S., 219
Enoki, T., 71

Félétou, M., 47
Fujita, H., 241
Fukui, T., 171
Fukuo, K., 113
Furchgott, R.F., 3
Furuta, H., 199

Garbers, D.L., 205
Gether, U., 191
Gibbons, G.H., 131
Griscavage, J.M., 13
Guo, D.-F., 199

Harpel, J., 151
Hashimoto, K., 21
Hattori, R., 21

Hayashi, Y., 171
Hirai, S., 171
Hirata, Y., 219

Ichikawa, A., 171
Ichiki, T., 199
Ignarro, L.J., 13
Imanaka, T., 253
Imura, H., 123
Inagami, T., 199
Inoue, T., 113
Ishikawa, M., 93
Ishikawa, T., 87
Itoh, H., 123
Iwai, N., 199

Jishage, K., 87
Jothianandan, D., 3

Kamada, N., 87
Kambayashi, Y., 199
Kamiya, A., 103
Kanaide, H., 61
Kaoru, M., 21
Karandikar, M., 211
Kawai, C., 21
Kawamoto, T., 21
Kawashima, H., 249
Kim, S., 93
Kodama, T., 87
Koizumi, T., 171
Korenaga, R., 103
Kosaka, C., 177
Kozaki, K., 93
Kumada, M., 87
Kurihara, H., 87
Kurihara, Y., 87

258

Kurokawa, K., 77
Kuwaki, T., 87

Lasky, L.A., 233

Maemura, K., 87
Masaki, T., 71
Masuda, J., 177
McEver, R.P., 227
Metz, C.N., 151
Miyasaka, M., 249
Miyazono, K., 157
Mori, M., 253
Morimoto, S., 113
Morita, I., 241
Morita, Y., 171
Murad, F., 29
Murota, S., 241

Nabata, T., 113
Nagai, R., 87
Nakahashi, T., 113
Nakajima, Y., 191
Nakanishi, N., 171
Nakanishi, S., 191
Nakao, K., 123
Nunes, I., 151

Oda, H., 87
Ogata, J., 177
Ogihara, T., 113
Ogoshi, S., 21
Ohyama, K., 199
Orimo, H., 93
Ouchi, Y., 87, 93

Pratt, R.E., 131

Rie-XiA Yang, 21
Rifkin, D.B., 151
Ross, E.M., 211

Ross, R., 143

Sakamoto, A., 71
Sasaguri, T., 177
Sasaki, K., 199
Sasayama, S., 21
Sase, K., 21
Schwartz, T.W., 191
Shima, K., 253
Shimokado, K., 177
Shizuta, Y., 21
Suzuki, H., 87

Takada, M., 205
Takahashi, K., 199
Takaku, F., 157
Takano, T., 253
Takemoto, F., 77
Tamatani, T., 249
Thompson, D.K., 205
Toba, K., 93
Toda, K., 21
Toda, N., 39
Toyoda, Y., 87

Uchida, S., 77

Vanhoutte, P.M., 47

Wakabayashi, Y., 241

Yamamoto, Y., 21
Yamano, Y., 199
Yazaki, Y., 87
Yokota, Y., 191
Yoshimasa, T., 123
Yoshitomi, K., 77
Yuen, P.S.T., 205
Yui, Y., 21

Zen, K., 177